Combinatorial Network Theory

Applied Optimization

Volume 1

The titles published in this series are listed at the end of this volume.

Combinatorial
Network Theory

Edited by

Ding-Zhu Du
University of Minnesota
and

D. Frank Hsu
Fordham University

KLUWER ACADEMIC PUBLISHERS
DORDRECHT / BOSTON / LONDON

Library of Congress Cataloging-in-Publication Data

Combinatorial network theory / edited by Din-Zhu Du and D. Frank Hsu.
    p.   cm. -- (Applied optimization ; vol. 1)

    1. Network analysis (Planning) 2. Combinatorial analysis.
I. Du, Dingzhu. II. Hsu, D. Frank (Derbiau Frank), 1948-
III. Series.
T57.85.C65 1995
511'.5--dc20
                                                        95-41542

ISBN 978-1-4419-4752-9

Published by Kluwer Academic Publishers,
P.O. Box 17, 3300 AA Dordrecht, The Netherlands.

Kluwer Academic Publishers incorporates
the publishing programmes of
D. Reidel, Martinus Nijhoff, Dr W. Junk and MTP Press.

Sold and distributed in the U.S.A. and Canada
by Kluwer Academic Publishers,
101 Philip Drive, Norwell, MA 02061, U.S.A.

In all other countries, sold and distributed
by Kluwer Academic Publishers Group,
P.O. Box 322, 3300 AH Dordrecht, The Netherlands.

Printed on acid-free paper

Printed in the Netherlands

CONTENTS

PREFACE

Recently the interest in the interconnection of communication has grown rapidly. One of the basic problems is to design optimal interconnection networks for certain needs. For example, to minimize the communication delay and to maximize the reliability, one looks for networks with minimum diameter and maximum connectivity under certain conditions. This small book consists of five chapters:

Chapter 1: Additive Group Theory Applied to Network Topology by Y.O. Hamidoune

Chapter 2, Connectivity of Cayley Digraphs by Ralph Tindell

Chapter 3, De Bruijn Digraphs, Kautz Digraphs, and Their Generalizations by Ding-Zhu Du, Feng Cao, and D. Frank Hsu

Chapter 4, Link-Connectivities of Extended Double Loop Networks by Frank K. Hwang and Wen-Ching Winnie Li

Chapter 5, Dissemination of Information in Interconnection Networks (Broadcasting & Gossiping) by Juraj Hromkovič, Ralf Klasing, Burkhard Monien, and Regine Peine

The subject of all of the chapters is the interconnection problem. The first two chapters deal with Cayley digraphs which are candidates for networks of maximum connectivity with given degree and number of nodes. The third chapter addresses de Bruijn digraphs, Kautz digraphs, and their generalizations, which are candidates for networks of minimum diameter and maximum connectivity with given degree and number of nodes. The fourth chapter studies double loop networks, and the fifth chapter considers broadcasting and gossiping problems. Each chapter may be read independently.

All chapters emphasize the combinatorial aspects of network theory. Combinatorial network theory aspires to two goals: solving practical problems and building up beautiful mathematics. We try to meet these two challenges in this book and hope to succeed with them.

Ding-Zhu Du
D. Frank Hsu

1

ADDITIVE GROUP THEORY APPLIED TO NETWORK TOPOLOGY

Y. O. Hamidoune
CNRS, Paris
France

1.1 INTRODUCTION

In this chapter we present some basic Addition theorems and their proofs. We explain how these results apply to network topology (connectivity, superconnectivity, girth and diameter). One of our goals is to bring powerful tools from Additive group theory to networks specialists.

A network will be modeled by a relation. Of course we will have always in mind the underlying graph. We will not use undirected graphs since they may be identified with symmetric graphs. This presentation seems most appropriate for the applications from group theory. Moreover it avoids a complicated terminology frequently used in graph theory.

Motivated by the famous Waring's problem, number theorists obtained some theorems expressing lower bounds on the size of the sum of two subsets in a group. The first result of this type is an inequality proved by Cauchy (1813) and rediscovered by Davenport (1935). It states that $|A + B| \geq \min(p, |A| + |B| - 1)$, where A and B are non empty subsets of $\mathbf{Z}/p\mathbf{Z}$, and p is a prime. The motivation of Cauchy was to prove that every element of $\mathbf{Z}_p$ is the sum of k kth powers. Davenport's motivation was to prove the p-analog of the famous $(\alpha+\beta)$ conjecture due to Khintchine (1932) and solved by Mann (1942). Mann's $(\alpha + \beta)$ Theorem can be formulated as follows. Let $A, B \subset \mathbf{N}$. Then $\sigma(A + B) \geq \min(1, \sigma(A) + \sigma(B))$, where $\sigma(A) = \inf\{|A \cap [1, n]|; n \in \mathbf{N}\}$. Note that this question raised in connection with a new proof of a theorem by Hilbert asserting that for any k there is g such that every integer is a sum of g kth powers.

1

Ding-Zhu Du and D. Frank Hsu (eds.), Combinatorial Network Theory, 1–39.
© *1996 Kluwer Academic Publishers. Printed in the Netherlands.*

In 1952, Mann proved the analog of his $(\alpha + \beta)$ theorem for any finite abelian group. A corollary of this result was applied to the Geometry of numbers by Kneser in 1955.

Several important contributions to Additive group theory have been obtained during the last 40 years. We mention few of them. A global Cauchy-Davenport inequality valid in any abelian group is proved by Kneser in 1955. Vosper obtained in 1956 a characterization of the cases of equality in the Cauchy-Davenport Theorem. Difficult results concerning the equality $|A + B| = |A| + |B| - 1$, where A and B are subsets of an abelian group were obtained by Kempermann (1961).

An independent work motivated by the study of the vulnerability of networks appeared recently in Combinatorics. This work is devoted to the study of the connectivity and the diameter of graphs with a transitive group of automorphisms.

It could be surprising that Additive group theory has many implications on these combinatorial problems. Actually all these questions for Cayley graphs have been more or less considered in Additive group theory. For this reason several facts from this theory have been rediscovered recently in connection with network vulnerability. The Cauchy-Davenport Theorem is rediscovered by the author (1977) in an equivalent form saying that the connectivity of a Cayley graph with a prime order is optimal. In 1970, Behzad and al conjectured that the minimum order of directed graph with girth g and degree d is $1 + d(g - 1)$. They were certainly unaware of a result of Shepherdson (1947) stating that this result holds for loop networks. In 1987, Boesch and Tindell rediscover a special case of the corollary of the finite $(\alpha + \beta)$ -Theorem mentioned above. The motivation of Boesch and Tindell was to obtain a necessary and sufficient condition for an undirected loop network to be with optimal connectivity.

Recently networks specialists become interested in constructing loop networks with given degree and minimum diameter. This question was considered in number theory in an equivalent formulation. It is known as the minimum base with a given order.

Our plan in this chapter is to introduce basic results from Additive group theory and to apply them to networks. Some of these applications are new. For example, the theory of atoms provides some properties of minimum cutsets. Unfortunately this theory does not apply to other cuts. We show here that such an information is contained in the finite $(\alpha + \beta)$ Theorem and Kneser's Theorem.

The diameter of the loop network defined by the k th powers was studied in number theory. We give here a nice method due to Chowla, Mann and Straus to prove that this number is bounded by $[k/2]+1$, in a prime order using Vosper's Theorem. The determination of superconnected loop networks, i.e. where every minimum cutset is of the $\partial^+(x)$ or $\partial^-(x)$ for some node x, received a lot of attention. We show how Vosper's Theorem implies easily that a loop network of a prime order not defined by an arithmetic progression is superconnected.

We prove Kneser's Theorem and mention some of its applications to abelian Cayley graphs. We give an outline of Kempermann critical pair theory and mention how to use it in order to characterize superconnected loop networks.

In the last section, we deal with the minimum cardinality of a base with given order and mention its connection with the minimum diameter of a loop network with a given degree.

1.2 BASIC NOTIONS

1.2.1 Preliminaries from Group Theory.

We shall assume some familiarity with the notion of a subgroup, of a coset and of a homomorphism. We use only elementary facts from group theory. We summarize them below for the commodity of the reader. We advice him to prove the following lemmas as exercises using only the definitions, which may be found in any book of group theory.

The cardinality of a set V will be denoted by $|V|$, if V is infinite we write $|V| = \infty$.

Let G be a group with a law written multiplicatively and let A and B be subsets of G. We write $A^{-1} = \{x^{-1} : x \in A\}$. The product of A and B is defined as $AB = \{xy : x \in A \text{ and } y \in B\}$. Let $x \in G$, we write xA for $\{x\}A$ and Ax for $A\{x\}$. In the case of abelian groups we will use the additive notations. In particular we write $A + B = \{x + y : x \in A \text{ and } y \in B\}$, etc.

Lemma 1.2.1 *Let G be a group containing three subsets A, B and C. Then $A(BC) = (AB)C$ and $(AB)^{-1} = B^{-1}A^{-1}$.*

The proof is immediate.

The *product* of A with itself k times will be denoted by A^k. In the additive notations, we use kA for $A + A + \cdots + A$, k times.

Let G be a group and let $x \in G$. The *left translation* with respect to x is the mapping $\gamma_x : G \longrightarrow G$, where $\gamma_x(y) = xy$, for any $y \in G$.

Lemma 1.2.2 *Let G be a group containing an element x and two subsets A and B. Then γ_x is a bijection from G onto G. Moreover $\gamma_x(A) = xA$. In particular $|xA| = |A|$, $x(A \cap B) = (xA) \cap (xB)$ and $x(A \backslash B) = (xA) \backslash (xB)$.*

The proof is immediate.

Lemma 1.2.3 *Let G be a group containing two non empty subsets A and B. Then $|AB| \geq |A|$. In particular $AB = A$ if and only if $AB \subset A$.*

The proof is immediate.

The above three lemmas will be applied without any reference.

Let G be a group containing an element x and a subset S and let H be a subgroup of G. We recall that xH is called a *left coset* of H. The *subgroup generated* by S is by definition the smallest subgroup of G containing S. We shall denote it by $< S >$.

We use the following lemmas.

Lemma 1.2.4 *Let G be a group containing a finite nonempty subset A. Then A is a subgroup if and only if $AA = A$.*

The proof is an easy exercise.

Lemma 1.2.5 *Let G be a group containing two finite subsets A and S. Then the following conditions are equivalent.*

(i) $AS = A$

(ii) $A < S >= A$

(iii) A is the union of left cosets of $< S >$.

The proof is left as an exercise.

We shall denote the ring of integers modulo n by $\mathbf{Z}_n$. The set of units of $\mathbf{Z}_n$ will be denoted by $\mathbf{Z}_n^*$.

Lemma 1.2.6 *Let $n \in \mathbf{N}$ and let $x \in \mathbf{Z}_n$. Then $x \in \mathbf{Z}_n^*$ if and only if $< x >= \mathbf{Z}_n$, where $< x >$ is the additive subgroup generated by x.*

Lemma 1.2.7 *Let p be a prime number. Then $\mathbf{Z}_p$ is a field. In particular, $\mathbf{Z}_p^*$ is a group with respect to multiplication.*

In the proof of Kneser's Theorem we will use the notion of factor group. We use the following well-known result.

Proposition 1.2.8 *Let f be an endomorphism of a finite group. Then $|Im(f)| \cdot |Ker(f)| = |G|$.*

In the last section, we shall use the structure of finite abelian groups. We summarize it below. Let G be an abelian group containing subgroups G_1, G_2, $\cdots$, G_k ($k \in \mathbf{N}$). Recall that G is said to be a *direct sum* of the family $\{G_1, G_2, \cdots, G_k\}$ if every element x of G has a unique expression $x = x_1 + x_2 + \cdots + x_k$, where $x_i \in G_i$, $1 \leq i \leq k$. This holds if and only if G is isomorphic to $G_1 \times G_2 \times \cdots \times G_k$. The following lemma is well-known.

Lemma 1.2.9 *Let $k \in \mathbf{N}$ and let G be an abelian group containing subgroups G_1, G_2, $\cdots$, G_k. Then G is a direct sum of $\{G_1, G_2, \cdots, G_k\}$ if and only if $G = G_1 + G_2 + \cdots + G_k$ and for every $1 \leq i \leq k-1$, $G_{i+1} \cap (G_1 + G_2 + \cdots + G_i) = \{0\}$.*

Let p be a prime number. A finite group G is said to be a p-group if there is $s \geq 1$ such that $|G| = p^s$. Let G be a finite abelian group. Following Lang, we write $A(p) = \{x : |< x >| \text{ is a power of } p\}$. The following result is well-known.

Theorem 1.2.10 *Let G be an abelian group and let $P = \{p : p$ is a prime and $A(p) \neq \{0\}\}$. For any $p \in P$, $A(p)$ is a p-group. Moreover G is a direct sum of the family $\{A(p); p \in P\}$.*

The structure of abelian p-groups is given by the following result.

Theorem 1.2.11 *Let G be an abelian p-group. Then G is a direct sum of cyclic p-groups.*

1.2.2   Networks Terminology

Let V be a set. The graph of the diagonal relation will be denoted by $\Delta(V)$. We have $\Delta(V) = \{(x, x) : x \in V\}$.

We recall some classical definitions from Set theory. A *relation* on a set V is an ordered pair (V, E), where E is a subset of $V \times V$. The set E will be called the *graph* of Γ. The elements of E will be called *arrows* or *arcs*. A relation will be sometimes called a *network*. In this case the points are usually called *nodes*. From now on we identify a relation with its graph and call the points vertices. The usual operations defined on relations (e.g., composition ...) will be applied to graphs.

Let Γ be a graph on a set V and let A be a subset of V. The *image* of A under Γ will be denoted by $\Gamma(A)$. We recall that

$$\Gamma(A) = \{y : \exists x \in A, (x, y) \in E\}.$$

We write $\partial_\Gamma(A) = \Gamma(A) \backslash A$. When the context is clear we write ∂ instead of ∂_Γ.

Let $x \in V$, we write $\Gamma(x) = \Gamma(\{x\})$ and $\partial(x) = \partial(\{x\})$. The *degree* of x is defined as $d_\Gamma(x) = |\partial(x)|$. If all points of V have the same degree, the graph will be called *regular*. Let Γ be a regular graph. The common degree of all points of Γ will be called the degree of Γ and will be denoted by $d(\Gamma)$. According to our definition, the degree of a point x with respect to $\Gamma \cup \Delta(V)$ is the same as $d_\Gamma(x)$. In others terms loops are not counted in the degree. However we will not exclude them in developing the theory. It is even desirable to add them in order to simplify some proofs and some notations.

Let $\Gamma = (V, E)$ be a graph. The *inverse* graph of Γ is the graph $\Gamma^- = (V, E^-)$, where $E^- = \{(x, y) : (y, x) \in E\}$.

A graph is said to be *locally finite* if every point has a finite degree. A graph will be called *finite* if its set of points is finite.

A graph Γ on a set V is said to be *connected* if for every $A \subset V$, such that $A \neq \phi$ and $A \neq V$, $\Gamma(A) \not\subset A$.

From now on, all graphs are assumed to be locally finite.

The *connectivity* of a graph $\Gamma = (V, E)$, denoted by $\kappa(\Gamma)$, is defined as follows.

$$\kappa(\Gamma) = min\{|\partial(F)| : |F| = 1 \ or \ \phi \neq F \cup \Gamma(F) \neq V\}.$$

Lemma 1.2.12 *Let* $\Gamma = (V, E)$ *be a graph. Then*

(i) $\kappa(\Gamma \cup \Delta(V)) = \kappa(\Gamma)$.

(ii) $\kappa(\Gamma) \leq min\{d(x) : x \in V\}$.

The proof is easy.

A subset F such that $\kappa(\Gamma) = |\partial(F)|$ and $F \cup \Gamma(F)$ is a proper subset of V is called a *fragment*.

Lemma 1.2.13 *Let* $\Gamma = (V, E)$ *be a graph. Then* $\kappa(\Gamma)$ *is the maximal integer k such that for any nonempty subset F of V,* $|F \cup \Gamma(F)| \geq min(|V|, |F| + k)$.

The proof follows easily from the definitions.

Lemma 1.2.14 *Let* $\Gamma = (V, E)$ *be a regular graph. Then the following conditions are equivalent.·*

(i) $\kappa(\Gamma) = d(\Gamma)$.

(ii) For any nonempty subset F of V,

$$|F \cup \Gamma(F)| \geq min(|V|, |F| + d(\Gamma)).$$

The proof is easy.

Lemma 1.2.15 *Let Γ be a finite graph. The following conditions are equivalent.*

(i) Γ is connected.

(ii) For any two points x and y, there is a path from x to y in the graph of Γ.

(iii) $\kappa(\Gamma) \geq 1$.

Let Γ be a graph on a set V and let $A \subset V$. The *subgraph* induced on A is by definition $\Gamma[A] = (A, E \cap (A \times A))$. A subset T is said to be a *cutset* of Γ if $\Gamma[V \backslash T]$ is not connected. It is clear that $\partial(F)$ is a cutset if $F \neq \phi$ and $F \cup \Gamma(F) \neq V$. Let A and T be nonempty subsets. We will say that T is a *fundamental cutset* with *support* A if $\partial(A) = T$ and $A \cup T \neq V$. The following lemma shows how to construct all cutsets from the fundamental cutsets.

Lemma 1.2.16 *Let $\Gamma = (V, E)$ be a graph and let $T \subset V$. Then T is a cutset of Γ if and only there is a fundamental cutset $T\prime \subset T$ and a support F of $T\prime$ such that $F \not\subset T$ and $(V \setminus (F \cup T\prime)) \not\subset T$.*

The proof is easy.

Lemma 1.2.17 *Suppose that $\Gamma \cup \Delta(V) \neq V \times V$. Then every minimal cutset is a fundamental cutset. Moreover $\kappa(\Gamma)$ is the minimal cardinality of a cutset of Γ.*

The proof is easy.

A cutset T is said to be a *minimum cutset* if $|T| = \kappa(\Gamma)$.

Lemma 1.2.18 ([13]) *Let Γ be a finite graph. Then $\kappa(\Gamma) = \kappa(\Gamma^{-})$.*

The proof is left to the reader.

Let $\Gamma = (V, E)$ be a regular finite graph. The graph Γ is said to be *superconnected* if $\kappa(\Gamma) = d(\Gamma)$ and if for every minimum cutset of Γ, there is $x \in V$ such that $T = \partial(x)$ or $T = \partial^{-}(x)$.

NOTICE. There is no standard terminology in Graph theory. The reader should take care before applying results from this area. The connectivity is called sometimes strong connectivity and in particular strongly connected should be replaced by connected. They use also the word outdegree for our notion of degree. In graph theory some authors use directed graph (or digraph) instead of graph. The definition of graph we use is contained in Bourbaki. It is standard in Mathematics, except some parts of graph theory.

A graph Γ is said to be *point-transitive* if its group of automorphisms acts transitively on its points. The following lemma is well-known and easy.

Lemma 1.2.19 *Let Γ be a point-transitive graph on a set V. Then Γ is regular. Moreover $d(\Gamma) = d(\Gamma^-)$, if V is finite.*

Let us now define the notion of *diameter* of a graph. We shall make this without reference to the notion of a path.

Let Γ be a graph on a set V. We recall that Γ^k is the relation obtained by composing Γ with itself k times. For any $x \in V$, we have $\Gamma^k(x) = \Gamma(\Gamma(\ldots(\Gamma(x))\cdots))$, k times. We write $\Gamma^0 = \Delta(V)$.

Let x and y be two points of V. The distance from x to y is by definition $dist_\Gamma(x,y) = \min\{k : y \in \Gamma^k(x)\}$, where $\min \phi = \infty$. The diameter of Γ is by definition $\rho(\Gamma) = \max\{dist_\Gamma(x,y) : x, y \in V\}$.

Notice that the the diameter is not affected by the addition of loops.

Lemma 1.2.20 *Let Γ be a graph on a set V and let x and y be two points of V. Then dist (x,y) is the minimal length of a path from x to y in the graph of Γ.*

The proof is easy.

Let $\Gamma = (V, E)$ be a graph. The *girth* of Γ is by definition $\eta(\Gamma) = 1 + \min\{dist(y,x) : (x,y) \in E \backslash \Delta(V)\}$.

Notice that the girth is not affected by the addition of loops.

Lemma 1.2.21 *Let Γ be a finite graph on a set V. Then $\eta(\Gamma)$ is the minimal length of a (directed) cycle in the graph of $\Gamma\backslash\Delta(V)$.*

The proof is easy.

We shall use these notions only in the case of point-transitive graphs. The following lemma shows that these notions become more simple in this case.

Lemma 1.2.22 *Let Γ be a point-transitive graph on a set V and let $v \in V$. Then*

(i) $\rho(\Gamma) = \max\{dist_\Gamma(v,x) : x \in V\}$.

(ii) $\eta(\Gamma) = \min\{k : v \in \Gamma^{k-1}(\partial(v))\}$.

The proof is easy.

1.2.3   Cayley Graphs

An important class of point-transitive graphs is the class of Cayley graphs defined below.

Let G be a group and let S be a subset of G. The *Cayley graph* on G defined by S is the graph $\Lambda(G,S) = (G,E)$, where $E = \{(x,y) : x^{-1}y \in S\}$.

Lemma 1.2.23 *Let G be a group containing a subset S and let $\Gamma = \Lambda(G,S)$ For every $F \subset G$, $\Gamma(F) = FS$ and $\partial(F) = (FS)\backslash F$.*

The proof follows from the definitions.

Lemma 1.2.24 *Let G be a group and let B be a finite subset of G. Then $\kappa(\Lambda(G,B))$ is the maximal integer k such that for any nonempty subset F of G,*

$$|F(B \cup \{1\})| \geq \min(|G|, |F| + k).$$

The proof follows from Lemmas 1.2.23 and 1.2.2.

Lemma 1.2.25 *Let G be a group and let B be a finite subset of G. Then*

$$\kappa(\Lambda(G, B \cup \{1\})) = |B \cup \{1\}| - 1$$

if and only if for all $A \neq \phi$,

$$|A(B \cup \{1\})| \geq \min(|G|, |A| + |B \cup \{1\}| - 1).$$

The proof follows from Lemmas 1.2.14 and 1.2.23.

Lemma 1.2.26 *Let G be a finite group and let $B \subset G$. Then*

(i) $\rho(\Lambda(G, B)) = \min\{k : (B \cup \{1\})^k = G\}.$

(ii) $\eta(\Lambda(G, B)) = \min\{k : 1 \in (B \setminus \{1\})(B \cup \{1\})^{k-1} = G\}.$

The proof follows from the definitions and Lemma 1.2.22

Lemma 1.2.27 *Let G be a finite group and let B be a subset of G. Then $\Lambda(G, B)$ is connected if and only if $G = < B >$.*

The proof is easy.

1.3   THE FINITE $(\alpha + \beta)$-THEOREMS

In subsection 1.3.1, we introduce the Dyson's transform and give its basic properties. We use it to prove the Cauchy-Davenport Theorem. This proof is closely related to that given in [46]. Note that several other proofs exist. In particular the reader may found an elegant and short proof in [8].

In subsection 1.3.2, we prove the Mann finite $(\alpha + \beta)$ Theorem and its additive consequences. The proof presented here is closely related to Mann's proof [36].

In subsection 1.3.3, we apply the finite $(\alpha + \beta)$-Theorem and its corollaries to the study of the cutsets in Cayley graphs. However, we think that this result can be used to prove other interesting properties of Cayley networks and advice the reader to search more applications for this result.

In subsection 1.3.4, we present two criteria due to Shepherdson and Scherk implying the Cauchy-Davenport inequality for a general abelian group. We present a corollary due to Shepherdon which determines the minimal order of Cayley graph with given degree and girth. The proofs given here for Shepherdon and Scherck are different from the orignal ones which use a trnsform due to Davenport.

1.3.1   The Cauchy-Davenport Theorem

The main result of this subsection is the Cauchy-Davenport Theorem. Let us define the Dyson's *transform*.

Let G be an abelian group containing two nonempty subsets A and B and let $e \in G$. The Dyson's e-transform of (A, B) is (A^*, B^*), where $A^* = A \cup (e + B)$ and $B^* = B \cap (A - e)$. The basic properties of the Dyson's transform are the following.

Lemma 1.3.1 *Let G be an abelian group containing two nonempty subsets A and B and let $e \in G$. Let $A^* = A \cup (e + B)$ and let $B^* = B \cap (A - e)$. Then*

(i) $A^* + B^* \subset A + B$.

(ii) $|A^*| + |B^*| = |A| + |B|$.

Proof. We have clearly

$$A^* + B^* = A \cup (e + B) + B \cap (A - e) \subset (A + B) \cup ((e + B) + (A - e)) \subset A + B.$$

We have, using Lemma 1.2.2,

$$|A^*| + |B^*| = |A \cup (e + B)| + |B \cap (A - e)|$$
$$= |A| + |e + B| - |A \cap (e + B)| + |B \cap (A - e)|$$
$$= |A| + |B|.$$

$\square$

Theorem 1.3.2 (Cauchy [5], Davenport [8]) *Let p be a prime number and let A, B be nonempty subsets of $\mathbf{Z}_p$. Then $|A + B| \geq \min(p, |A| + |B| - 1)$.*

Proof. The proof is by induction on $|B|$. The result is obvious for $|B| = 1$. Suppose the result is proved for all B' with $|B'| < k$, where $k = |B| \geq 2$. Take $b \in B$ and put $S = B - b$. We have clearly $0 \in S$ and $< S >= \mathbf{Z}_p$.

Let A be a subset of G. The inequality of the theorem holds trivially if $|A| = 1$. Assume $|A| \geq 2$. Suppose first that for every $a \in A$, $a + S \subset A$. Therefore $A + S \subset A$. By Lemma 1.2.3, $A + S = A$. By Lemma 1.2.6, $|A| \geq |< S >| = p$. In this case the inequality of the theorem is obviously verified.

Suppose now that there exists $a \in A$ such that $(a + S)\backslash A \neq \phi$. Put $A^* = A \cup a + S$ and $S^* = S \cap (A - a)$. Since $(a + S) \not\subset A$, we have using Lemma 1.2.3, $|S^*| < |S|$. Clearly $0 \in S$. By the induction hypothesis we have

$$|A^* + S^*| \geq \min(p, |A^*|+|S^*| - 1).$$

By Lemma 1.2.12, we have $|A + S| \geq \min(p, |A|+|S|-1)$. But clearly $|S| = |B|$ and $|A + S| = |A + B|$. This last remark completes the proof of the theorem. $\square$

Theorem 1.3.2 is the p-analog of the $(\alpha + \beta)$-Theorem. According to Lemma 1.2.24, Theorem 1.3.2 is equivalent to the following result rediscovered in 1977.

Theorem 1.3.3 ([13]) *Let p be a prime number and let B be a nonempty subset of $\mathbf{Z}_p$. Then $\kappa(\Lambda(\mathbf{Z}_p, B)) = |B\backslash\{0\}|$.*

1.3.2   Mann's Theorem

The following lemma is well-known.

Lemma 1.3.4 *Let A, B be finite subsets of a group G such that $AB \neq G$. Then $|A| + |B| \leq |G|$.*

Proof. Let $x \in G\backslash AB$. We have clearly $xB^{-1} \cap A = \phi$. It follows

$$|A| + |B| = |A|+|B^{-1}x| = |A \cup B^{-1}x| \leq |G|. \quad \square$$

Proposition 1.3.5 (Mann, [34, 36]) *Let G be a finite abelian group and let A, B be subsets of G such that $0 \notin A + B$. There is a subgroup H of G and a subset W with the following properties.*

(i) $A \subset W$

(ii) $W + B = G \backslash H$

(iii) $|W| - |A| = |W + B| - |A + B|$.

Proof. The proof is by induction on $k(A, B) = |G| - |A + B|$. If $k(A, B) = 1$, the theorem holds trivially with $W = A$ and $H = \{0\}$. Suppose $k \geq 2$. Put $D = G \backslash (A + B)$. Since $0 \in D$, we have clearly $D \subset D + D$. If $D + D = D$ then by Lemma 1.2.5, D is a subgroup, in this case the theorem holds trivially with $W = A$ and $H = D$. Suppose the contrary and choose $(r, s, a, b) \in D \times D \times A \times B$ such that $r + s = a + b$. Put $A^* = (a - D) \cap (D - B)$ and $C = A \cup A^*$.

We begin by showing the equality

$$(A^* + B) \cap D = D \cap (a - A^*) \tag{1.1}$$

We have clearly $a - A^* = D \cap (a + B - D)$. Let $x \in (A^* + B) \cap D$. Choose $(v, u) \in B \times A^*$ such that $x = u + v$. Since $u \in (a - D)$, we have $x \in a + B - D$. This shows the first inclusion.

Let $x \in (a - A^*) \cap D = D \cap (a + B - D)$. Choose $v \in B$ and $y \in D$ such that $x = a + v - y$. Now $a - y = x - v \in D - B$. Clearly $a - y \in (a - D)$. Therefore $a - y \in A^*$. It follows that $x \in A^* + B$. This shows the second inclusion.

Let us show that $A^* \cap A = \phi$. Suppose on the contrary that there exists $u \in A \cap A^*$. It follows that $u \in A \cap (D - B)$, a contradiction. From (1), we get $|(A^* + B) \cap D| = |D \cap (a - A^*)|$. Hence

$$|(A^* + B) \cap D| = |(a - D) \cap A^*| = |(a - D) \cap (D - B)| = |A^*|. \tag{1.2}$$

Therefore

$$|C| - |A| = |A^*| = |C + B| - |A + B| \tag{1.3}$$

Now we have clearly $a - r \in A^*$. It follows that $a - r + b \in (A^* + B) \backslash (A + B)$. Therefore $|G| - |A + C| < |G| - |A + B|$. We have also $0 \notin C + B$, since otherwise there is $t \in B$ and $q \in D$, such that $q - r \in A$, a contradiction.

By the induction hypothesis, there exists a subgroup H and a subset W such that

$$C \subset W, \ C + W = G \backslash H \text{ and } |W| - |C| = |C + W| - |C + B| \tag{1.4}$$

By (1.3) and (1.4), we have $|W| - |A| = |W + B| - |A + B|$, which proves the theorem. $\qquad\square$

The condition $0 \notin A + B$ may be replaced by $x \notin A + B$, where x is an arbitrary element of G. However the general case follows from the case $0 \notin A + B$.

Theorem 1.3.6 (Mann, [34, 36]) *Let G be a finite abelian group and let A, B be subsets of G such that $c \notin A + B$. There is a subgroup H of G and a subset W with the following properties.*

(i) $A \subset W$ and $W + B = G \backslash (c + H)$

(ii) $|W| - |A| = |W + B| - |A + B|$.

Proof. Let $B' = B - c$. Clearly $0 \notin A + B'$. By Theorem 1.3.4, there are a subgroup H and a subset W such that

$$A \subset W \text{ and } W + B' = G \backslash H \qquad (i)$$

$$|W| - |A| = |W + B'| - |A + B'|. \qquad (ii)$$

The result follows now since $W + B = W + B' + c = G \backslash (c + H)$, $|W + B| = |W + B'|$ and $|A + B| = |A + B'|$. $\qquad\square$

We will refer to Theorem 1.3.6 as the finite $(\alpha + \beta)$–Theorem.

Corollary 1.3.7 (Mann, [34, 36]) *Let G be a finite abelian group and let A, B be nonempty subsets of G such that $A + B \neq G$. There exists a subgroup H such that $H + B \neq G$ and $|A + B| \geq |A| + |B + H| - |H|$.*

Proof. Suppose $A + B \neq G$ and choose $x \in G \backslash (A + B)$. By Theorem 1.3.6, there are a subgroup H and a subset $W \supset A$ with the following property.

$$W + B = G \backslash (x + H) \text{ and } |A + B| - |A| = |W + B| - |W|. \qquad (1.5)$$

Clearly, $|W + B| = |G| - |x + H| = |G| - |H|$. Since $W + H + B = W + B + H = (G \backslash (x + H)) + H \neq G$, we have by Lemma 1.3.4, $|G| \geq |W| + |B + H|$. It follows from (1) that $|A + B| - |A| \geq |W + B| - |W| \geq |B + H| - |H|$.

Corollary 1.3.8 (Mann, [34, 36]) *Let G be a finite abelian group and let B be a subset of G. Suppose that for every subgroup H of G, $|H + B| \geq \min(|G|, |H| + |B| - 1)$. Then for every nonempty subset A, $|A + B| \geq \min(|G|, |A| + |B| - 1)$.*

Proof. Suppose there is $A \neq \phi$ such that $|A + B| < \min(|G|, |A| + |B| - 1)$. It follows that $A + B \neq G$. By Corollary 1.3.7, there is a subgroup H such that $H + B \neq G$ and $|A + B| \geq |A| + |B + H| - |H|$. Hence $|H + B| < |G|$. Also we have $|B| - 1 > |A + B| - |A| \geq |B + H| - |H|$. It follows that $|H + B| < \min(|G|, |H| + |B| - 1)$. □

Corollary 1.3.9 (Chowla [6]) *) Let n be a natural number and let A, B be nonempty subsets of $\mathbf{Z}_n$ such that $0 \in B$ and $B \backslash \{0\} \subset (\mathbf{Z}_n)^*$. Then $|A + B| \geq \min(n, |A| + |B| - 1)$.*

Proof. Assume $|A + B| < n$. By Corollary 1.3.7, there is a subgroup H such that $|H + B| < n$ and $|A + B| \geq |A| + |B + H| - |H|$. Therefore $|H| < n$. By Lemma 1.2.7, $H \cap B = \{0\}$, since any element of $B \backslash \{0\}$ generates $\mathbf{Z}_n$. Clearly $H + B \supset H \cup B$. It follows that $|H + B| \geq |H \cup B| \geq |H| + |B| - 1$. Hence $|A + B| \geq |A| + |B + H| - |H| \geq |A| + |B| - 1$. □

Obviously Chowla's Theorem implies the Cauchy-Davenport Theorem.

Exercise: Adapt the proof in subsection 1.3.1 for the Cauchy-Davenport Theorem to obtain a proof of Chowla's Theorem.

1.3.3   Cutsets in Cayley Graphs

Several results about the connectivity of abelian Cayley graphs follow easily from the finite $(\alpha + \beta)$−Theorem of Mann. This result can be formulated in the language of graphs as follows.

Theorem 1.3.10 (The $(\alpha + \beta)$−Theorem for networks) *Let Γ be a connected Cayley graph defined on a finite abelian group G and let T be a fundamental cutset of Γ with support A. Let $c \in G \backslash (A \cup T)$. There is a subgroup H of G and a fundamental cutset $T\prime$ with a support W such that $W \supset A$ $W \cup T\prime = G \backslash (c + H)$ and $|T| = |T\prime|$.*

Proof. Put $\Gamma = \Lambda(G, S)$, where $S \subset G$. Put $B = S \cup \{0\}$. By Lemma 1.2.23, $c \notin A + B$. By Theorem 1.3.2, there are a subgroup H and a subset $W \supset A$ such that $A \subset W$, $W + B = G \backslash (c + H)$ and $|W| - |A| = |W + B| - |A + B|$. By the last equality we have $|W| - |A| = |W \cup W + S| - |A \cup A + S|$. By Lemma 1.2.23, we have $|\partial(W)| = |\partial(A)| = |T|$. □

Corollary 1.3.11 *Let Γ be a connected Cayley graph defined on a finite abelian group G and let T be a fundamental cutset of G. There is a subgroup H of G such that $H \cup \Gamma(H) \neq G$ and $|\partial(H)| \leq |T|$.*

The proof is left as an exercise.

Corollary 1.3.12 *Let Γ be a Cayley graph defined on a finite abelian group Then $\kappa(\Gamma) = d(\Gamma)$ if and only if for any subgroup H, $|H \cup \Gamma(H)| \geq \min(|G|, |H| + d(\Gamma))$.*

The reader may prove easily that this result is equivalent to Corollary 1.3.8.

A strictly equivalent formulation of a this result applied to symmetric graphs is the following result rediscovered in 1987.

Theorem 1.3.13 (Boesch-Tindell) *Let n be an integer and let $S \subset \mathbf{Z}_n \backslash \{0\}$ such that $S = -S$. Then $\kappa(\Lambda(\mathbf{Z}_n, S)) = |S|$ if and only if for any proper divisor of m the number of positive residues of the elements of S modulo m is at least $\min(m, |S|m/n)$.*

The equivalence follows by substituting in Corollary 1.3.12 the value of the subgroups of $\mathbf{Z}_n$. □

1.3.4 Subsets with the Cauchy-Davenport Inequality

Chowla's Theorem gives a condition on B asserting the validity of the Cauchy-Davenport inequality for an arbitrary $A \subset \mathbf{Z}_n$. The following criterion applies to all finite abelian groups. We present here a proof based on the Dyson's transform. Note that Shepherdson's proof is elegant and short. However it uses the Davenport's transform.

Theorem 1.3.14 (Shepherdson [43]) *Let G be an abelian group and let A, B be two subsets of G such that $0 \in B$ and $B \subset A \cup \{0\}$. Then one of the following conditions holds.*

(i) There is $b \in B \backslash \{0\}$ such that $< b > \subset A + B$

(ii) $|A + B| \geq |A| + |B| - 1$.

Proof. The proof is by induction on $|B|$. The result is obvious for $|B| = 1$. Suppose it proved for all the subsets with cardinality strictly less than $|B|$. Let $a \in B \backslash \{0\}$. Assume $< a > \not\subset (A + B)$. Since $a \in A$, there is $k \geq 1$ such that $ka \in A$ and $(k + 1)a \notin A$. Otherwise $< a > \subset A \subset A + B$. Set $ka = r$. We have $B \not\subset A - r$, since otherwise $(k + 1)a \in A$. Therefore, $B \cap (A - r)$ is a strict subset of B. By the induction hypothesis, there exists $x \in B \cap (A - r)$ such that $< x > \subset (A \cup r + B) + B \cap (A - r)$ or $|(A \cup r + B) + B \cap (A - r)| \geq |A \cup r + B| + |B \cap (A - r)| - 1$. By Lemma 1.3.1, either $< x > \subset A + B$ or $|A + B| \geq |A| + |B| - 1$. □

Corollary 1.3.15 (Shepherdson [43]) *Let G be an abelian group, S be a subset of G such that $0 \notin S$ and let k be an integer such that $k|S| \geq |G|$. Then $0 \in S \cup 2S \cup \cdots \cup kS$.*

Proof. Suppose on the contrary that $0 \notin (S \cup 2S \cup \cdots \cup kS)$. Hence

$$|S \cup 2S \cup \cdots \cup kS| < |G| \leq k|S|.$$

Let j be the smallest i such that

$$|S \cup 2S \cup \cdots \cup iS| < i|S|.$$

Clearly $2 \leq j \leq k$. By the definition of j, we have

$$|S \cup 2S \cup \cdots \cup (j - 1)S| \geq (j - 1)|S|.$$

Set $B = S \cup \{0\}$. Since $0 \notin S \cup 2S \cup \cdots \cup jS = (S \cup 2S \cup \cdots \cup (j-1)S) + B$, we have by Theorem 1.3.14, $|S \cup 2S \cup \cdots \cup jS| \geq |S \cup 2S \cup \cdots \cup (j - 1)S| + |S| \geq j|S|$. This contradiction proves the corollary. □

We shall formulate now Corollary 1.3.15 using the notion of girth.

Corollary 1.3.16 *Let G be a finite abelian group and let Γ be a Cayley graph on G. Then $|G| \geq d(\Gamma)(\eta(\Gamma) - 1) + 1$.*

Proof. Set $\Gamma = \Lambda(G, S)$. We may assume without lost of generality $0 \notin S$, by the definition of the girth given subsection 1.2.2. Put $k = \lceil |G|/|S| \rceil$. We have clearly $k|S| \geq |G|$. By Corollary 1.5.9, $0 \in S \cup 2S \cdots \cup kS$. By Lemma 1.2.25, $\eta(\Gamma) \leq k \leq (|G| + |S| - 1)/|S|$. Therefore $|G| \geq d(\Gamma)(\eta(\Gamma) - 1) + 1$. $\square$

The reader may verify easily that Corollary 1.3.15 is equivalent to Corollary 1.3.16.

The validity of Corllary 1.3.16 for any regular finite graph has been conjectured by Behzad and al [2]. They were certainly unaware of Shepherdson's result implying its validity for the most important class of regular graphs.

This conjecture is proved for point transitive graphs.

Theorem 1.3.17 (Hamidoune [15]) *Let Γ be a point-transitive graph on a finite set V. Then $|V| \geq 1 + d(\Gamma)(\eta(\Gamma) - 1)$.*

The proof uses the theory of atoms. At that moment we were unaware of Shepherdson's theorems.

We give now an other useful criterion for the validity of the Cauchy-Davenport inequality. This result was conjectured by Moser and proved by Scherk. We present here a proof based on the Dyson's transform. Note that Scherk's proof is elegant and short. However it uses the Davenport's transform.

Theorem 1.3.18 (Scherck) *Let A and B be finite subsets of an abelian group. Suppose that $0 \in A \cap B$ and $0 \notin A + (B \setminus \{0\})$. Then $|A + B| \geq |A| + |B| - 1$.*

Proof. The proof is by induction on $|B|$, the result being obvious for $|B| = 1$. We may assume $|B| \geq 2$ and let $b \in B \setminus \{0\}$. We show first that there exists $x \in A$ such that $B \not\subset (A - x)$. Suppose the contrary. It follows that $A + B = A$. By Lemma 1.2.17, A is a union of $< B >$-cosets. Since $0 \in A$, we have $< B > \subset A$, and hence $-b \in A$. Therefore $0 = -b + b \in A + (B \setminus \{0\})$, a contradiction. Choose $a \in A$ such that $B \not\subset (A - a)$. It follows that $|(A - a) \cap B| < |B|$. We have clearly $0 \in B \cap (A - a)$. Let us verify that $0 \notin (A \cup a + B)(B \cap (A - a) \setminus \{0\})$.

Suppose the contrary. There exists $x \in A \backslash a$ and $y \in B$ such that $(a+y)+(x-a) = 0$. By the hypothesis of the theorem, we have $x = 0$ and $y = 0$. Therefore $-a \in B$, and hence $0 = a + (-a)$. It follows that $a = 0$ and hence $x = a, a$ contradiction. By Lemma 1.3.1 and the induction hypothesis, we have

$$
\begin{aligned}
|A+B| &\geq |(A \cup a + B) + (B \cap A - a)| \\
&\geq |A \cup a + B| + |B \cap A - a| - 1 \\
&\geq |A| + |B| - 1.
\end{aligned}
$$

$\square$

Corllary 1.3.16 (and also Theorem 1.3.17 applied to abelian Cayley graphs) is rediscovered by Alon who gave a short proof based on Theorem 1.3.18. He applies it combined with a Ramsey argument to a number theory conjecture of Erdös and Graham.

Exercise: Deduce Corollary 1.3.16 frome Theorem 1.3.18.

1.4   THE CRITICAL PAIR PROBLEM

1.4.1   The e-Transform

The following transform will be used in subsection 1.4.2. It is one of the basic tools in both Additive group theory and Additive number theory.

Let G be an abelian group containing two nonempty subsets A and B and let $c \in G \backslash (A+B)$. Set $D = G \backslash (A+B)$. For any $e \in B$, the (c,e)-*transform* of B is $B(e,c) = B \cup (e+c-D) \backslash B$.

The basic properties of the e-transform are given in the following lemma.

Lemma 1.4.1 *Let G be an abelian group containing two nonempty subsets A and B. Set $D = G \backslash (A+B)$ and let $c \in D$.*

(i) $c \notin A + B(e,c)$

(ii) $D \cap (A + B(e,c)) \subset c + e - (B(e,c) \backslash B)$

(iii) $|A + B(e,c)| - |B(e,c)| \leq |A+B| - |B|$.

Proof. Suppose $c \in A + B(e, c)$. There are $a \in A$ and $u \notin A + B$ such that $c = a + e + c - u$. It follows that $u \in A + B$, a contradiction, which proves (i). Let $x \in D \cap (A + B(e, c))$. There is $y \in A$ and $z \notin A + B$ such that $x = y + c + e - z$. Therefore $z = y + c + e - x$. It follows that $c + e - x \in (B(e, c) \backslash B)$. Hence $x \in c + e - (B(e, c) \backslash B)$. This proves (ii). We have using (ii)

$$|(A + B(e, c)) \backslash (A + B)| = |D \cap (A + B(e, c))|$$
$$\leq \quad |c + e - (B(e, c) \backslash B)| = |B(e, c) \backslash B|$$

$\square$

1.4.2 Vosper's Theorem

We begin by two lemmas.

Lemma 1.4.2 *Let p be a prime number and let A be a nonempty subset of $\mathbf{Z}_p$. Let $d \in \mathbf{Z}_p^*$. If A is not an arithmetic progression then there are $x, y \in A$ such that $x \neq y$, $x + d \notin A$ and $y + d \notin A$.*

The proof is easy.

Lemma 1.4.3 *Let p be a prime number and let A be a subset of $\mathbf{Z}_p$ such that $|A| \geq 2$. Let $B = \{b, b + d, \cdots, b + kd\}$, where $b, d \in \mathbf{Z}_p$ and $k \geq 1$ is a natural number. Then either $|A + B| \geq \min(p, |A| + |B|)$, or A is an arithmetic progression with difference d.*

Proof. By considering the subsets $A' = d^{-1}(A - b)$ and $B' = d^{-1}(B - b)$, we reduce the problem to the case where $B = \{0, 1, \cdots, k\}$. The lemma becomes an easy exercise. $\square$

The crucial part of Vosper's critical pair Theorem is the following statement which has interesting implications for loop networks superconnectivity. The general statement will be given at the end of the subsection.

Theorem 1.4.4 (Vosper) *Let p be a prime number and let A be a subset of $\mathbf{Z}_p$ which is not an arithmetic progression. For any subset B such that $|B| \geq 2$, $|A + B| \geq \min(p - 1, |A| + |B|)$.*

Proof. The proof is by induction on $k(A, B) = p - |A + B|$. The result being obvious for $k(A, B) = 1$, since $|A + B| = p-1$ in this case. Set $D = \mathbf{Z}_p \backslash (A+B)$. Assume $k(A, B) \geq 2$ and let $c \in D$. Suppose the result proved for all B' such that $k(A, B') < k(A, B)$. The result holds by Lemma 1.4.3 if B is an arithmetic progression, suppose the contrary. We consider two cases.

Case 1. There is $e \in B$ such that $(e+c-D)\backslash B \neq \phi$ and $(e+c-D) \cap (B \backslash \{e\}) \neq \phi$

Clearly $|B(e, c)| \geq |B| + 1$. We shall show that $|A + B(e, c)| \leq p - 2$. Take $u \in D$ such that $e + c - u \in B \backslash \{e\}$. We have $u \notin A + B(e, c)$, since otherwise $u = a + e + c - v$, for some $a \in A$ and $v \notin A + B$, this would imply $v \in A + B$, a contradiction. But $u \neq c$. By Lemma 1.4.1, $\{u, c\} \cap (A + B(e, c)) = \phi$. Hence

$$|A + B(e, c)| \leq p - 2. \tag{1.6}$$

Suppose $|A + B| < |A + B(e, c)|$, we have $k(A, B(e, c)) < k(A, B)$. By the induction hypothesis, $|A + B(e, c)| \geq \min(p - 1, |A| + |B(e, c)|)$. Using (1), we have $|A + B(e, c)| \geq |A| + |B(e, c)|$. By Lemma 1.4.1, $|A + B| \geq |A + (B(e, c)| - |B(e, c) \backslash B| + |B| \geq |A| + |B|$. Therefore $|A + B| \geq |A| + |B|$. Assume now $A + B = A + B(e, c)$, we have by the Cauchy-Davenport Theorem, Lemma 1.4.1 and (1.6), $|A + B| \geq |A| + |B(e, c)| - 1 \geq |A| + |B|$.

Case 2. For every $e \in B$ either

$$(e + c - D) \cap B = \{e\} \quad \text{or} \quad e + c - D \subset B. \tag{1.7}$$

It follows that for every $e \in B$, either $B(e, c) = B$ or $B(e, c) \backslash B = (e + c - D) \backslash \{e\}$.

Let $w \in D \backslash \{c\}$. By Lemma 1.4.2, there are $x, y \in B$ such that $x \neq y$ and $x + c - w, y + c - w \notin B$. It follows that $x + c - w \in B(x, c) \backslash B$. Using (1.7), we have

$$B(x, c) \backslash B = (x + c - D) \backslash \{x\}. \tag{1.8}$$

Similarly

$$B(y, c) \backslash B = (y + c - D) \backslash \{y\} \tag{1.9}$$

$$\text{Take } B_0 = (B(x, c) \cup B(y, c)) \backslash B.$$

By Lemma 1.4.1,

$$|A + B| \geq |A + (B \cup B_0)| - |B_0|. \tag{1.10}$$

We may assume $k(A, B \cup B_0) < k(A, B)$, since otherwise $A + B = A + (B \cup B_0)$. By The Cauchy-Davenport Theorem, $|A + B| \geq |A| + |B| + |B_0| - 1 \geq |A| + |B|$. By the induction hypothesis,

$$|A + (B \cup B_0)| \geq \min(p - 1, |A| + |B| + |B_0|). \tag{1.11}$$

Assume first $|A + (B \cup B_0)| \leq p - 2$. We have by (5) and (6), $|A + B| \geq |A| + |B|$. It remains to consider the case $|A + (B \cup B_0)| = p - 1$. By (5), $|A + B| \geq p - 1 - |B_0|$. It follows that $|D| - 1 \leq |B_0|$. By the definitions we have, $x + c - (D\backslash\{c\}) = y + c - (D\backslash\{c\})$. It follows that $(D\backslash\{c\}) + y - x = (D\backslash\{c\})$. By Lemma 1.2.17, $D\backslash\{c\}$ is a union of $< y - x >$-cosets. Since $< y - x >= \mathbf{Z}_p$, we have $D\backslash\{c\} = \phi$ and hence $k(A, B) = 1$, a contradiction.

The Cauchy-Davenport Theorem asserts that $|A + B| \geq \min(p, |A| + |B| - 1)$ for all nonempty subsets A and B. The critical pair problem for this inequality consists in characterizing all the pairs $\{A, B\}$ such that $p - 1 \geq |A + B| = |A| + |B| - 1$. Let us mention some cases where the double inequality holds.

(1) $|A| = 1$ and $|B| < p$ or $|B| = 1$ and $|A| < p$.

(2) A and B are arithmetic progressions with the same difference and $|A| + |B| \leq$p.

(3) There is $x \in \mathbf{Z}_p$ such that $A = \mathbf{Z}_p\backslash(x - B)$.

The equality holds trivially in (1) and (2). Let us verify it in (3). Suppose $A = \mathbf{Z}_p\backslash(x - B)$. By the Cauchy-Davenport Theorem we have $|A + B| \geq \min(p, |A| + |B| - 1) = p - 1$. On the other side $x \notin A + B$. Therefore $|A + B| = p - 1 = |A| + |B| - 1$. □

The following result of Vosper shows that there are no other cases where equality holds.

Theorem 1.4.5 (Vosper [47, 48]) *Let p be a prime number and let A and B be subsets of $\mathbf{Z}_p$ such that $|A| \geq 2$ and $|B| \geq 2$. Then one of the following conditions holds true.*

(1) $|A + B| \geq \min(p, |A| + |B|)$.

(2) There is $x \in \mathbf{Z}_p$ such that $A = \mathbf{Z}_p\backslash(x - B)$.

(3) A and B are arithmetic progressions with the same difference.

Proof. The result holds by Lemma 1.4.3, if B is an arithmetic progression. Suppose that B is not an arithmetic progression. By Theorem 1.4.4, $|A + B| \geq \min(p - 1, |A| + |B|)$. Then (1) holds unless $p - 1 = |A + B| \leq |A| + |B| - 1$.

By the Cauchy-Davenport Theorem $|A + B| = |A| + |B| - 1 = p - 1$. Let $\{x\} = \mathbf{Z}_p \backslash (A + B)$. We have $A \subset \mathbf{Z}_p \backslash (x - B)$, since otherwise $x \in A + B$. On the other side we have $|A| = p - |B| = |\mathbf{Z}_p \backslash (x - B)|$. Therefore $A = \mathbf{Z}_p \backslash (x - B)$ and hence (2) is satisfied.   □

Exercise. Characterize all the pairs $\{A, B\}$ such that $p = |A + B| = |A| + |B| - 1$.

1.4.3 The Waring's Graph

We begin by estimating the size of jP, where P is the set of k th powers in $\mathbf{Z}_p$. We recall that $jP = P + P + \cdots + P$, j times.

Lemma 1.4.6 ([36]) *Let p be a prime, k, n be integers and let $P = \{x^k : x \in \mathbf{Z}_p\}$. Then $|nP| \equiv 1 \pmod{|P| - 1}$.*

Proof. Let $U = P \backslash \{0\}$. It is clear that U is multiplicative subgroup of $\mathbf{Z}_p^*$. We have clearly

$$U(nP \backslash \{0\}) = U((P + P + \cdots + P) \backslash \{0\}) = (UP + UP + \cdots + UP) \backslash \{0\} = nP \backslash \{0\}.$$

By Lemma 1.2.17, $nP \backslash \{0\}$ is a union of cosets of U. Therefore

$$|nP| - 1 \equiv 0 \pmod{|U|}.$$

□

Lemma 1.4.7 (Lemma 2.1.2 in [36]) *Let p be a prime and let U be a multiplicative subgroup of $\mathbf{Z}_p^*$ such that $3 \le |U| < p - 1$. Then $U \cup \{0\}$, is not an arithmetic progression.*

Proof. The hypothesis implies $p \ge 5$. In order to formalize the proof we shall denote the unity in $\mathbf{Z}_p$ by ϵ. Let us first prove that $\sum_{x \in W} x = 0$ for any non null subgroup W of $\mathbf{Z}_p^*$, $\sum_{x \in W} x = 0$. Take $w \in W \backslash \{1\}$. We have clearly $\sum_{x \in W} wx = \sum_{x \in W} x = w \sum_{x \in W} x$. Hence $(1 - w) \sum_{x \in W} x = 0$. The hypothesis imply that both U and $\{x^2 : x \in U\}$ are non null. Therefore

$$\sum_{x \in U} x = \sum_{x \in U} x^2 = 0.$$

Suppose now that $U \cup \{0\}$ is an arithmetic progression. Say $U \cup \{0\} = \{-jd, -(j-1)d, \cdots, 0, d, \ldots, hd\}$, where $h + j = |U|$. Clearly $d \in U$. It follows that $d^{-1}U = U$. Hence $U \cup \{0\} = \{-j\epsilon, -(j-1)\epsilon, \cdots, 0, \epsilon, \ldots, h\epsilon\}$. By (1) $j = h$. Therefore we have, using the above equality, $\sum_{x \in U} s^2 \epsilon = 0$ and hence $(j+1)j(2j+1)\epsilon = 0$. Since $|U| \leq (p-1)/2, 2j+1 < p$. Therefore $\epsilon = 0$, a contradiction. $\qquad\qquad\square$

Theorem 1.4.8 (Chowla-Mann-Straus [7]) *Let p be a prime, let k, n be integers and let $P = \{x^k : x \in \mathbf{Z}_p\}$. If $p - 1 \geq |P| \geq 4$, then $|nP| \geq \min(p, (2n-1)(|P|-1)+1)$.*

Proof. The result is obvious for $P = \mathbf{Z}_p$, suppose $|P| \leq p - 1$. Since $P\backslash\{0\}$ is a subgroup of $\mathbf{Z}_p^*$, we have $|P\backslash\{0\}| \leq (p-1)/2$. By Lemma 1.4.7, P is not an arithmetic progression. The proof is by induction. The statement is obvious for $n = 1$. Suppose it true for n. We shall show that

$$|(n+1)P| \geq \min(p, (2n+1)|P| + 1).$$

Assume $|(n+1)P| \leq p - 1$. Since $|P| - 1$ divides $p - 1$, $|P| - 1$ is not a divisor of $p - 2$. By Lemma 1.4.6, $|P| - 1$ divides $|(n+1)P| - 1$. Therefore $|(n+1)P| \leq p-2$. By Theorem 1.4.4, we have $|(n+1)P| \geq \min(p-1, |nP|+|P|)$ Since $p - 2 \geq |(n+1)P| \geq |nP|$, we have by the induction hypothesis

$$|(n+1)P| \geq (2n-1)(|P|-1) + 1 + |P| = 2n(|P|-1) + 2.$$

By Lemma 1.4.6,
$$|(n+1)P| \equiv 1 \pmod{(|P|-1)}.$$

Therefore
$$|(n+1)P| \geq (2n+1)(|P|-1) + 1.$$

$\qquad\qquad\square$

Let us now apply these results to determine the diameter of the Waring's graph $\Lambda(\mathbf{Z}_p, U)$. Notice that this graph has interesting symmetry properties. In particular it is transitive on the arcs.

Lemma 1.4.9 *Let p be a prime number and let k be a divisor of $p - 1$. Then $|\{x^k : x \in \mathbf{Z}_p^*\}| = (p-1)/k$.*

Proof. Consider the endomorphism ν of $\mathbf{Z}_p^*$, where $\nu(x) = x^k$. The condition on k shows that $|Ker(\nu)| = k$. By Proposition 1.2.21,

$$|Im(\nu)| = (p-1)/|Ker(\nu)| = (p-1)/k.$$

Therefore

$$|\{x^k : x \in \mathbf{Z}_p^*\}| = |Im(\nu)| = (p-1)/k.$$

$\square$

Corollary 1.4.10 *Suppose k divides $p-1$ and $k \neq (p-1)/2$ and let $P = \{x^k : x \in \mathbf{Z}_p\}$. Then $\rho(\Lambda(\mathbf{Z}_p, U)) \leq [k/2] + 1$.*

Proof. We have by Lemma 1.4.9 and Theorem 1.4.8,

$$|([k/2]+1)P| \geq \min(p, (2[k/2]+1)(p-1)/k + 1) = p.$$

By Lemma 1.2.26,

$$\rho(\Lambda(\mathbf{Z}_p, U)) \leq [k/2] + 1.$$

$\square$

1.4.4   Superconnectivity in Loop Networks

The problem of determining superconnected loop networks received a lot of attention in network theory. However no general result has been obtained before the use of Additive group theory. The first result of this type is the application of Vosper's Theorem to obtain the following.

Proposition 1.4.11 (Hamidoune-Tindell [25]) *Let p be a prime number and let $S \subset \mathbf{Z}_p^*$. If $S \cup \{0\}$ is not an arithmetic progression, then $\Lambda(\mathbf{Z}_p, S)$ is superconnected.*

Proof. Let $B = S \cup \{0\}$. We have $\kappa(\Lambda(\mathbf{Z}_p, S)) = |S|$, by the Cauchy-Davenport Theorem. Let T be cutset of $\Gamma = \Lambda(\mathbf{Z}_p, S)$ and let F be a subset such that $\partial(F) = (F + S) \backslash F = T$ and $F \cup T \neq \mathbf{Z}_p$. We have clearly $|F + B| = |F \cup T| = |F| + |B| - 1$. Since B is not an arithmetic progression we have either $|F| = 1$ or $F = \mathbf{Z}_p \backslash (x - B)$. It follows that for some x either $F = \{x\}$ or $F = \mathbf{Z}_p \backslash (\{x\} \cup \Gamma^-(x))$. The reader may verify easily that $\partial(\mathbf{Z}_p \backslash (\{x\} \cup \Gamma^-(x))) = \partial^-(x)$. Therefore $T = \partial(x)$ or $T = \partial^-(x)$. Hence Γ is superconnected.

1.4.5 Kempermann Theory

Let G be an abelian group. We recall that a subset K is called periodic if there is a non null subgroup H such that $K = K + H$.

Following Kempermann a pair $\{A, B\}$ of nonempty subsets of G is called an elementary pair if one of the conditions (i)-(iv) holds true.

(i) $|A| = 1$ or $|B| = 1$.

(ii) A and B are arithmetic progressions with difference d, where d is of order $\geq |A| + |B| - 1$.

(iii) For some finite subgroup H each of A and B is contained in a H-coset while $|A| + |B| = |H| + 1$.

(iv) B is aperiodic and for some finite subgroup H of G, B is contained in a H-coset while A is of the form $g - ((B + H) \setminus B)$, where $g \in G$.

The main structure theorem of Kempermann is the following.

Theorem 1.4.12 (Kemperman [28]) *Let G be an abelian group such that $|G| \geq 2$ containing two nonempty finite subsets A and B such either $A + B$ is aperiodic or there exist $a \in A$ and $c \in A + B$ such that $c \notin (A \setminus a) + B$. Then a necessary and sufficient condition in order that $|A + B| = |A| + |B| - 1$ is the existence of a nonempty subset A_1 of A and a nonempty subset B_1 of B and a subgroup F with $|F| \geq 2$ satisfying the following conditions.*

(I) The pair $\{A_1, B_1\}$ is elementary and each of A_1 and B_1 is contained in some F-coset.

(II)$(A_1 + B_1) \cap ((A \setminus A_1) + B) = \phi$ and $(A_1 + B_1) \cap (A + (B \setminus B_1)) = \phi$.

(III) Both $A \setminus A_1$ and $B \setminus B_1$ are a union of F-cosets.

(IV) $|\sigma(A) + \sigma(B)| = |\sigma(A)| + |\sigma(B)| - 1$, where σ is the canonical morphism from G onto G/F.

The purpose of Kempermann's theory is to determine all the pairs $\{A, B\}$ where $|A + B| \leq |A| + |B| - 1$ and $A + B \neq G$.

Theorem 1.4.12 is used in [23] to characterize superconnected Cayley graphs on abelian groups.

1.5   KNESER'S THEOREM AND SOME APPLICATIONS

1.5.1   Kneser's Theorem

We use the following lemmas.

Lemma 1.5.1 *Let G be a direct sum of two subgroups X and Y. Let U and V be two subsets of G such that $U + X = U$ and $V + Y = V$. Then $|U \cap V| = |U||V|/|X||Y|$.*

Proof. Take $|U + X| = u|X|$ and $|V + Y| = v|Y|$. By Lemma 1.2.5, $u, v \in \mathbf{N}$. Let $A \subset U$ be such that $|A| = u$ and $A + X = U$ and let $B \subset V$ be such that $|B| = v$ and $B + Y = V$. We have clearly $U \cap V = \bigcup_{a \in A, b \in B}(a + X) \cap (b + Y)$.

Since $X \cap Y = \{0\}$, one verifies easily that $|(a + X) \cap (b + Y)| \leq 1$. Therefore $|U \cap V| \leq uv$.

Let us show first that $(a + X) \cap (b + Y) \neq \phi$, for every $(a, b) \in A \times B$. Take $a - b = x + y$, with $x \in X$ and $y \in Y$. We have clearly $a - x = b + y \in (a + X) \cap (b + Y)$. Suppose now $a \neq a'$. We have $(a + X) \cap (a' + X) = \phi$, since A is a transversal. It follows that $|U \cap V| = |A||B| = uv$.   □

Lemma 1.5.2 *Let G be group containing two subgroups X and Y such that $X \cap Y = \{0\}$. Let U, V be two subsets of G such that $U + X = U$ and $V + Y = V$. Suppose $U \backslash V \neq \phi$ and $V \backslash U \neq \phi$. Then either $|U \backslash V| \geq |Y| - 1$ or $|V \backslash U| \geq |X| - 1$.*

Proof. Set $|Y| = y$ and $|X| = x$. The proof will be by contradiction. Suppose

$$|U \backslash V| \leq y - 2 \quad and \quad |V \backslash U| \leq x - 2 \qquad (1.12)$$

Choose $e \in U \backslash V$. Set $P = (U - e) \cap (X + Y)$ and $Q = (V - e) \cap (X + Y)$. Since U and $X + Y$ are union of X- cosets, so is P, i.e. $P + X = P$. Similarly

$Q + Y = Q$. Take $|P| = p|X|$ and $|Q| = q|Y|$. We have $Q\backslash P = (V - e) \cap (X + Y)\backslash(U - e) \cap (X + Y) = ((V\backslash U) \cap (X + Y + e)) - e$. Therefore $|Q\backslash P| \le |(V\backslash U) \cap (X + Y)| \le |V\backslash U|$. Similarly $|P\backslash Q| \le |U\backslash V|$. By (1), we have

$$|P\backslash Q| \le y - 2 \text{ and } |Q\backslash P| \le x - 2. \tag{1.13}$$

By Lemma 1.2.9, $X + Y$ is a direct sum of X and Y. By Lemma 1.5.1, we have $|P \cap Q| = pq$. Therefore $|P\backslash Q| = |P| - |P \cap Q| = p(x - q)$ and similarly $|Q\backslash P| = q(y - p)$.

Using (2) we have $x + y - 4 \ge |P\backslash Q| + |Q\backslash P| = p(x - q) + q(y - p)$. Therefore

$$2pq - 4 \ge (p - 1)x + (q - 1)y. \tag{1.14}$$

Since $e \in U\backslash V$, we have $0 \in P\backslash Q$. Therefore $x \ge q + 1$. By (1.13), $y - 2 \ge |P\backslash Q| = p(x - q) \ge p$. Therefore $y \ge p + 2$. By (1.14) and the inequalities $x \ge q + 1$ and $y \ge p + 2$, we have Hence $2pq - 4 \ge x(p - 1) + y(q - 1) \ge (q + 1)(p - 1) + (p + 2)(q - 1) = 2pq + q - 3$, a contradiction. $\square$

Proposition 1.5.3 *Let G be an abelian group containing two subgroups X and Y. Let U, V be two subsets of G such that $U + X = U$ and $V + Y = V$. Suppose $U\backslash V \ne \phi$ and $V\backslash U \ne \phi$. Then either $|U\backslash V| \ge |Y| - |Y \cap X|$ or $|V\backslash U| \ge |X| - |Y \cap X|$.*

Proof. Set $H = X \cap Y$. If $H = \{0\}$, the result follows by Lemma 1.5.2. Suppose $|H| \ge 2$. Consider the canonical mapping $\nu : G \longrightarrow G/H$. Since $X \supset H$ and $Y \supset H, \nu(X)$ and $\nu(Y)$ are subgroups of G/H. Since ν is a homomorphism, we have $\nu(U) + \nu(X) = \nu(U)$ and $\nu(V) + \nu(Y) = \nu(V)$. Let us prove that $\nu(U\backslash V) = \nu(U)\backslash\nu(V)$. We have $\nu(U\backslash V) = \nu(\nu^{-1}(\nu(U))\backslash\nu(\nu^{-1}(\nu(V))) = \nu(\nu^{-1}(\nu(U)\backslash\nu(V))) = (\nu(U)\backslash\nu(V))$. Similary $\nu(V\backslash U) = \nu(V)\backslash\nu(U)$. It follows that $\nu(U)\backslash\nu(V) \ne \phi$ and $\nu(V)\backslash\nu(U) \ne \phi$. Clearly $\nu(X) \cap \nu(Y) = \{0\}$. By Lemma 1.5.2, either $|\nu(U)\backslash\nu(V)| \ge |\nu(Y)| - 1$ or $|\nu(V)\backslash\nu(U)| \ge |\nu(X)| - 1$. The result follows now since $|U\backslash V| = |\nu(U)\backslash\nu(V)||H|$ and $|V\backslash V| = |\nu(V)\backslash\nu(U)||H|$. $\square$

Lemma 1.5.4 *Let G be an abelian group containing two finite subsets A and B. Suppose $0 \in B$ and let $c \in A$ There is a subset C such that*

$$c \in C = C + H \subset A + B \text{ and } |C| \ge |A| + |B| - |H|.$$

Proof. The proof is by induction on $|B|$. The result is obvious for $|B| = 1$. If $A + B = A$, then by Lemma 1.5.1. $A+ < B > = A$. The result holds clearly with $C = A$ and $H = < B >$. Suppose $A + B \neq A$. Choose $e \in A$ such that $e + B \not\subset A$. Consider the Dyson's transform $A^* = A \cup e + B$ and $B^* = B \cap (A - e)$. We have $0 \in B^*$ and $c \in A^*$. We have also $|B^*| \leq |B| - 1$, since otherwise we would have $e + B \subset A$. By the induction hypothesis, there exist a subgroup H^* and a subset C^* such that $c \in C^* = C^* + H^* \subset A^* + B^*$ and $|C^*| \geq |A^*| + |B^*| - |H^*|$. By Lemma 1.2.2, the result holds with $C = C^*$ and $H = H^*$. $\square$

Proposition 1.5.5 (Kneser) *Let G be an abelian group containing two finite subsets A and B. For every subset $T \subset A + B$, there is $C \subset A + B$ such that*

$$T \subset C = C + H \subset A + B \text{ and } |C| \geq |A| + |B| - |H|.$$

Proof. The proof is by induction on $|T|$. Let us prove it for $|T| = 1$. Set $T = \{t$ and choose $a \in A$ and $b \in B$. By Lemma 1.5.4 there is a subgroup H and a subset C_0 such that

$$a \in C_0 = C_0 + H \subset A + B - b \text{ and } |C_0| \geq |A| + |B| - |H|.$$

The result holds with $C = C_0 + b$. Assume $k = |T| \geq 2$. Suppose the result true for all subsets with cardinality less than k. Let $t \in T$. By the induction hypothesis there are a subgroup H_1 and a subset C_1 such that

$$T \backslash \{t\} \subset C_1 = C_1 + H_1 \subset A + B \text{ and } |C_1| \geq |A| + |B| - |H_1|. \tag{1.15}$$

By Lemma 1.5.4, there exist a subgroup H_2 and a subset C_2 such that

$$t \in C_2 = C_2 + H_2 \subset A + B \text{ and } |C_2| \geq |A| + |B| - |H_2|. \tag{1.16}$$

If $C_1 \subset C_2$ or $C_2 \subset C_1$, the result is obvious. Suppose the contrary. By Proposition 1.5.3, either

$$|C_1 \backslash C_2| \geq |H_2| - |H_1 \cap H_2| \text{ or } |C_2 \backslash C_1| \geq |H_1| - |H_1 \cap H_2|. \tag{1.17}$$

$$\text{Put } C = C_1 \cup C_2 \text{ and } H = H_1 \cap H_2.$$

We have clearly $C = C_1 \cup (C_2 \backslash C_1) = C_2 \cup (C_1 \backslash C_2)$. It follows easily using (1.15), (1.16) and (1.17) that

$$|C| \geq |A| + |B| - |H_1 \cap H_2|.$$

We have also $C + (H_1 \cap H_2) = C_1 + (H_1 \cap H_2) \cup C_2 + (H_1 \cap H_2) = C_1 \cup C_2$. $\square$

The following form of Kneser's theorem is the most standard one. The reader may check easily that it implies Proposition 1.5.5.

Theorem 1.5.6 (Kneser) *Let G be an abelian group containing two finite subsets A and B. There exists a finite subgroup K such that $A+B+K = A+B$ and $|A + B| \geq |A + K| + |B + K| - |K|$*

Proof. Let K be the maximal subgroup such that $A + B + K = A + B$. By Proposition 1.5.5, applied with to $A + K$ and $B + K$ and $C = A + K + B + K = A + B$, there exists a subgroup H such that $A + B + K + H = A + B + K$ and $|A + B + K| \geq |A + K| + |B + K| - |H|$. It follows by the maximality of K that $K + H = K$ and therfore $H \subset K$. Therefore $|A + B| = |A + B + K| \geq |A + K| + |B + K| - |K|$. □

The following equivalent form of Kneser's Theorem is quite appropriate for some applications. It was observed by Kempermann in [28].

Theorem 1.5.7 (Kneser) *Let G be an abelian group containing two finite subsets A and B such that $|A + B| \leq |A| + |B| - 2$. There exists a finite subgroup H such that $A + B + H = A + B$ and $|A + B| = |A + H| + |B + H| - |H|$.*

Proof. By Theorem 1.5.6, there is a subgroup H such that $|A + B + H| \geq |A + H| + |B + H| - |H|$. Since $|A + B| \leq |A| + |B| - 2$, we have $|H| \geq 2$. On the other side we have $|A + B + H| \leq |A + H| + |B + H| - 2$. Observe that all the terms are multiples of $|H|$, except 2. Therefore $|A + B| = |A + B + H| \leq |A + H| + |B + H| - |H|$. □

The reader may easily verify that Theorem 1.5.7 and Theorem 1.5.6 are in fact equivalent.

1.5.2 Applications to Cutsets

We use now Kneser's theorem to study the structure of minimal cutsets. Let G be an abelian group containing a subgroup H and let A be a subset of G. We recall that A is said to be H-periodic if A is a union of H-cosets. By Lemma 1.2.5, A is H-periodic if and only if $A + H = A$.

Proposition 1.5.8 *Let G be an abelian group and let Γ be a Cayley graph on G. Let T be a minimal cutset of Γ such that $|T| \leq d(\Gamma) - 1$. There exists a non null subgroup H such that $H \cup \Gamma(H) \neq G$ and $|T| = |\partial(H)|$. Moreover T*

is H-periodic and therefore $(|G|, |T|) \geq 2$. Let A be a proper sink of $\Gamma[G \setminus T]$. Then A is $H-$ periodic.

Proof. Take $\Gamma = \Lambda(G, S)$ and put $B = S \cup \{0\}$. Clearly $T = \partial(A)$ and $A \cup T \neq G$. By Theorem 1.5.7, there is a subgroup H such that

$$|A + B + H| = |A + H| + |B + H| - |H| \text{ and } A + B = A + B + H.$$

But $\partial(A + H) = (A + H + S) \setminus (A + H) = (A + H + B) \setminus (A + H) \subset (A + B) \setminus A \subset \partial(A) = T$. Since $(A + H) \cup \Gamma(A + H) = A + B = A \cup \Gamma(A) \neq G, \partial(A + H)$ is a cutset. By the minimality of T we have $T = \partial(A + H) = (A + H + B) \setminus (A + H)$. Therefore T is H-periodic. In particular $|H|$ divides $|T|$ and hence $(|G|, |T|) \geq 2$. We have also $(A + H + B) \setminus (A + H) = A + B \setminus A$. It folllows that $(A + H) \subset A$. Hence A+H=A.   □

Corollary 1.5.9 *Let G be an abelian group and let Γ be a symmetric Cayley graph on G. Let T be a minimal cutset of Γ such that $|T| \leq d(\Gamma) - 1$. Let C be a connected component of $\Gamma[G \setminus T]$. There exists a non null subgroup H such that A is $H-$ periodic.*

Proof. Clearly C is a sink.   □

1.6   BASES OF FINITE ABELIAN GROUPS

1.6.1   Rohrbach Problem

Let G be a finite group and let $h \in \mathbf{N}$. A subset $B \subset G$ is said to be a h-base if $B^h = G$.

Lemma 1.6.1 *Let G be a finite group and let $h \in \mathbf{N}$. A subset B containg 1 is a h-base if and only if $\rho(\Lambda(G, B)) \leq h$.*

Rohrbach [38, 39] asked if for every $h \geq 2$ there exists c such every finite group G has a base of order $c|G|^{1/h}$.

Recently network theorists became interested in constructing loop networks with given order and degree and with minimum diameter. The reader may found some details about the work done on this question in network theory in the survey paper [3]. This problem is only a reformulation of Rohrbach's question.

1.6.2   Bases in Abelian Groups

Let H be a subgroup of G. A subset $T \subset G$ will be called a *transversal* of G/H if T intersects each $H-$ coset in exactly one element. The results presented here are contained in [27]. We shall use the following lemmas. Let G be a group containing a subgroup

Lemma 1.6.2 *[27] Let p be a prime number and let h, m be two natural numbers. Let G be a finite abelian group with order p^{mh}. There are subsets $A_1, A_2, \cdots, A_h$ such that $|A_1| = |A_2| = \cdots = |A_h| = m$ and $A_1 + A_2 + \cdots + A_h = G$.*

Proof. It is well-known (cf. [32]) that there is a tower of subgroups $\{0\} = G_0 \subset G_1 \subset \cdots \subset G_h = G$ and $|G_i| = p^{im}$. Let A_i be a transversal of G_{i+1}/G_i, $0 \le i \le h-1$. Clearly $|A_1| = |A_2| = \cdots = |A_h| = m$ and $A_1 + A_2 + \cdots + A_h = G$. $\square$

Lemma 1.6.3 *Let h and n be two natural numbers. There are subsets A_1, A_2, $\cdots$, A_h such that $|A_1| = |A_2| = \cdots = |A_h| = \lceil n^{1/h} \rceil$ and $A_1 + A_2 + \cdots + A_h = \mathbf{Z}_n$.*

Proof. Put $u = \lceil n^{1/h} \rceil$. Set $B_i = u^{i-1}\{0, 1, \cdots, u-1\}, 1 \le i \le h$. Observe that $B_i \subset \mathbf{N}$. We have clearly

$$B_1 + B_2 + \cdots + B_h = \{a_0 + a_1 u + \cdots + a_{h-1} u^{h-1} : 0 \le a_i \le u - 1\}.$$

Therefore,

$$B_1 + B_2 + \cdots + B_h \supset [0, u^h - 1].$$

It follows that

$$B_1 + B_2 + \cdots + B_h \supset [0, n - 1].$$

Let ν be the canonical morphism from $\mathbf{N}$ onto $\mathbf{Z}_n$. Put $A_i = \nu(B_i)$. We have $A_1 + A_2 + \cdots + A_h = \nu(B_1 + B_2 + \cdots + B_h) \supset \nu[0, n-1] = \mathbf{Z}_n$. On the other side, one verifies easily that

$$|A_1| = |A_2| = \cdots = |A_h| = u = \lceil n^{1/h} \rceil.$$

$\square$

Lemma 1.6.4 *Let $S \subset \mathbf{N}$ and let h be an non zero integer. There exists $T \subset S$ such that $|S \backslash T| \leq h - 1$ and $\sum_{i \in T} i \equiv 0 \pmod{h}$.*

Proof. Let T be a maximal subset of S with sum $\equiv 0$, mod h. Put $S \backslash T = \{t_1, t_2, \cdots, t_k\}$. Consider the mapping $\psi(i) = t_1 + t_2 + \cdot, s + t_i \bmod h, 1 \leq i \leq k$. We have $\psi(i) \neq \psi(j)$, for $i \neq j$, since otherwise T would not be maximal. It follows that $k \leq h - 1$. $\square$

Lemma 1.6.5 *Let p be a prime number and let h, s be a natural numbers. Let G be a finite abelian group with order p^s. There are an integer m and cyclic subgroups $G_2, \cdots, G_h$ and a subgroup G_1 such that $|G_1| = p^{mh}$ and $G = G_1 \oplus G_2 \oplus \cdots \oplus G_h$.*

Proof. By Theorem 1.2.10, G is a direct sum of cyclic subgroups $\{H_i; i \in I\}$. By Lemma 1.6.4, there is $J \subset I$ such that $\sum_{j \in J} |H_j| = mh$, for some m and $|I \backslash J| \leq h - 1$. We choose G_1 to be the direct sum of the subgroups $\{H_i; i \in J\}$. By the associativity and commutativity of the direct sum, G is the direct sum of G_1 and $\{H_i; i \in I \backslash J\}$. It follows that G is the direct sum of G_1 and at most $h - 1$ cyclic subgroups. By adding null subgroups, we obtain the result. $\square$

Lemma 1.6.6 *Let h be a natural number and let G be a finite abelian group. There are an integer m and cyclic subgroups $G_2, \cdots, G_h$ and a subgroup G_1 such that $|G_1| = m^h$ and $G = G_1 \oplus G_2 \oplus \cdots \oplus G_h$.*

Proof. Let P be the set of primes dividing $|G|$. By Theorem 1.2.10, G is a direct sum of $\{H(p); p \in P\}$, where $H(p)$ is a p-subgroup. By Lemma 1.6.6, $H(p) = G_{p1} \oplus G_{p2} \oplus \cdots \oplus G_{ph}$, where $|G_{p1}| = m_p^h$ and $G_{p2}, \cdots, G_{ph}$ are cyclic subgroups. Take $K(j) = \oplus_{p \in P} G_{pj}; 1 \leq j \leq h$. Clearly $K(j)$ is a cyclic subgroup. Obviously $|K(1)| = \pi_{p \in P}(m_{p1})^h$. The Lemma follows now easily. $\square$

Lemma 1.6.7 *Let h and j be natural numbers such that $j \geq 2$. Let $a_1, a_2,, a_h$ be non zero natural numbers . Then*

$$\lceil a_1^{1/h} \rceil \lceil a_2^{1/h} \rceil \cdots \lceil a_j^{1/h} \rceil < (1 + 2^{-j})^{(j-1)} (a_1 a_2 \cdots a_j)^{1/h}.$$

The proof is an easy exercise.

Proposition 1.6.8 *[27] Let h be a natural number and let G be a finite abelian group. There are subsets $A_1, A_2, \quad \cdots, A_h$ such that $|A_1| = |A_2| = \cdots = |A_h| = \lceil |G|^{1/h} \rceil$ and $A_1 + A_2 + \cdots + A_h = G$.*

Proof. By Lemma 1.6.6, there are an integer m and subgroups cyclic subgroups $G_2, \cdots, G_h$ and a subgroup G_1 such that $|G_1| = m^h$ and $G = G_1 \oplus G_2 \oplus \cdots \oplus G_h$. By Lemma 1.6.3, there are subsets $A_{11}, A_{12}, \quad \cdots, A_{1h}$ such that $|A_{11}| = |A_{12}| = \cdots = |A_{1h}| = m$ and $A_{11} + A_{12} + \cdots + A_{1h} = G_1$. Take $i \geq 2$. By Lemma 1.6.3, there are subsets $A_{i1}, A_{i2}, \cdots, A_{ih}$ such that $|A_{i1}| = |A_{i2}| = \cdots = |A_{ih}| = \lceil |G_i|^{1/h} \rceil$ and $A_{i1} + A_{i2} + \cdots + A_{ih} = G_i$. Put $A_i = A_{1i} + A_{2i} + \ldots + A_{hi}$. We have $A_1 + A_2 + \cdots + A_h = \sum_i A_{i1} + A_{i2} + \cdots A_{ih} = G_1 + G_2 + \cdots + G_h = G$. Put $a_i = \lceil |G_i|^{1/h} \rceil$. Clearly $|A_i| = a_1 a_2 \cdots a_h$ By Lemma 1.6.8, $|A_i| < (1 + 2^{-h})^{h-1} |G|^{1/h}$. $\qquad\qquad\square$

Theorem 1.6.9 *[27] Let h be a natural number and let G be a finite abelian group. There exists a base of G with order h and cardinality at most $h(1 + 2^{-h})^{h-1} |G|^{1/h}$.*

Take B to be the union of the subsets given by Proposition 1.6.8. $\qquad\qquad\square$

1.6.3   The Plünnecke Inequalities

Let G be an abelian group containing a subset A. Recall the notation $jA = \{x_1 + x_2 + \cdots x_j \big| x_i \in A\}$.

The following nice inequalities, which may give relations between the cardinality of balls in abalian Cayley graphs where proved by Plünnecke in Additive number Theory.

Theorem 1.6.10 (Plünnecke, [33]) *Let A be a finite subset of an abelian group G and let $k \leq j$. Then*

$$|kA| \geq |jA|^{\frac{k}{j}}.$$

The proof of Plünnecke was simplified by Rusza in [40].

It is known that Cayley graphs Cayley graphs on abelian has poor expansion properties [1] Annextsein and Baumslag [1] showed that abelian Cayley graphs have small cuts separating the graph into big components.

This result was improved in [24] as follows.

Theorem 1.6.11 *[24] Let G be an abelian group of order n and let S a generating subset of G such that $|S| \leq n - 2$. Let ρ be the diameter of $\Gamma = \Lambda(G, S)$.*

There exists a cutset of size less than $\frac{4n \ln(n/2)}{\rho}$ whose deletion separates Γ into a sink B and a source $\overline{B}$ such that $|B| = |\overline{B}|$.

Corollary 1.6.12 *[24] Let Γ be a connected Cayley graph on an abelian group with oudegree r .*

There exists a cutset of size $< \frac{8e}{r} n^{1-1/r} \ln(n/2)$ whose deletion separates Γ into a sink B (which is a ball) and a source $\overline{B}$ such that $|B| = |\overline{B}|$.

REFERENCES

[1] F. Annexstein, M. Baumslag, On the diameter and bisector size of Cayley graphs, Mathematical Systems Theory , 271-291 (1993).

[2] M. Behzad, G. Chartrand and C. Wall, On minimal regular graphs with given girth, Fund. Math. 69 (1970), 227-231.

[3] J.C. Bermond, F. Comellas and D.F. Hsu, Distributed loop computer networks: a survey, preprint (1990),

[4] F. Boesch and R. Tindell, Circulants and their Connectivities, J. Graph Theory, 8 (1984) 487-499.

[5] A. Cauchy, Recherches sur les nombres, J. Ecole polytechnique 9(1813), 99-116.

[6]I. Chowla, A theorem on the addition of residue classes: Applications to the number $\Gamma(k)$ in Waring's problem, Proc. Indian Acad. Sc. 2 (1935), 242-243.

[7] S. Chowla, H.B. Mann and E.G. Straus, Some applications of the Cauchy-Davenport theorem, Norske Vid. Selsk. Forh. (Trondheim) 32(1959), 74-80.

[8] H. Davenport, On the addition of residue classes, J. London Math. Society10(1935), 30-32.

[9] H. Davenport, A historical note, J. London Math. Society 22(1947), 100-101.

[10] C. Delorme and Y.O. Hamidoune, On the product of subsets in groups Graphs and Combinatorics, Graphs and Combinatorics 10 (1994), 101 - 104.

[11] M.M. Dodson and A. Tietäväinen, A note on Waring's problem in GF(p), Acta Arithmetica XXX(1976), 159-167.

[12] J. Fàbrega and M.A. Fiol, Maximally connected digraphs, J. Graph Theory 13(1989), 657-668.

[13] Y.O. Hamidoune, Sur les atomes d'un graphe orienté, C.R. Acad. Sc. Paris A 284 (1977), 1253-1256.

[14] Y.O. Hamidoune, Quelques problèmes de connexité dans les graphes orientés, J. Comb. Theory B, 30 (1981), 1-10.

[15] Y.O. Hamidoune, An application of connectivity theory in graphs to factorization of elements in groups, Europ. J of Combinatorics, 2 (1981), 108-112.

[16] Y.O. Hamidoune, On the connectivity of Cayley digraphs, Europ. J. Combinatorics 5 (1984), 309-312.

[17] Y.O. Hamidoune, Sur la séparation dans les graphes de Cayley Abeliens, Discrete Math. 55(1985),323-326.

[18] Y.O. Hamidoune, Sur les atomes des graphes de Cayley infini, Discrete Math., 73 (1989), 297-300.

[19] Y.O. Hamidoune, Factorisations courtes dans les groupes finis, Discrete App. Math., 24 (1989), 153-165.

[20] Y.O. Hamidoune, A note on the addition of residues, Graphs and Combinatorics 6(1990), 147-152.

[21]Y.O. Hamidoune, On some graphic aspectes of Addition Theorems, Ringel Feitzeitschrift, Topics in Combinatorics and Graph Theory, Physica-Verlag Heidelberg (1990), 311-318.

[22] Y.O. Hamidoune, On subsets with small sums in abelian groups, J. Number Theory, soumis.

[23] Y.O. Hamidoune, A. S. Llàdo and O. Serra, Vosperian and superconnected Abelian Cayley digraphs, Graphs and Combinatorics 7(1991), 143-152.

[24] Y.O. Hamidoune and O. Serra, On small cuts separating an abelian Cayley graph into two equal parts, Mathematical Systems Theory, to appear

[25] Y. O. Hamidoune and R. Tindell, Some applications of additive group theory to connectivity, manuscript (1989).

[26] H. Heilbronn, Lecture notes on additive number theory mod p, California Institute of Technology, 1964.

[27]X.D. JIA, Thin bases in finite abelian groups, J. Numb. Theory 36 (1990), 254-256.

[28]J.H.B. Kempermann, On small sumsets in Abelian groups, Acta Math. 103 (1960), 66-88.

[29] A Khintchine, Zur Additiven Zahlentheorie, Mat. Sbornik 39 (1932), 27-34.

[30] M. Kneser, Anwendung eines satzes von Mann auf die Geometrie of Zahlen, Proceedings Int. Cong. Math. Amsterdam 2 (1954), 32.

[31]M. Kneser, Abschäntzungen der asymptotischen Dichte von summenmengen, Math. Zeit. 58(1953), 459-484.

[32] S. Lang, Algebra, Adison Wesley Publ. Company, 1978.

[33] H. Plünnecke, Eine Zahlentheoretische Anwendung der Graphentheorie, J. Reine Angew. Math. 243 (1970), 171-183.

[34] H.B. Mann, A proof of the fundamental theorem on the density of sums of sets of positive integers, Ann. Math. 43 (1942), 523-527.

[35] H.B. Mann, An addition theorem of sets of elements of an abelian group, Proc. Amer. Math. Soc. 4 (1953), 4.

[36]H.B. Mann, Addition theorems : The addition theorems of group theory and number theory, Interscience, New York, 1965.

[37] J.E. Olson, On the sum of two sets in a group, J. Number Theory 18(1984), 110-120.

[38]H. Rohrbach, Eine Beitrag zur additiven Zahlentheorie, Math. Z. 42 (1937), 1-30.

[39] H. Rohrbach, Anwendung eine Satzes der additiven Zahlentheories auf eine gruppen-theorisches Frage, Math. Z. 42 (1937), 537-542.

[40]I.Z. Rusza, An application of Graph theory to additive number theory, Scientia Ser. A 3 (1989), 97-109.

[41] S. Schwarz, On Waring's problem for finite fields, Quart. J. Math. Oxford 19 (1948), 123-128.

[42]S.Schwarz, On equations in finite fields, Quart. J. Math. Oxford 19 (1948), 160-163.

[43] J.C. Shepherdson, on the addition of elements of a sequence, J. London Math. Soc. 22(1947), 85-88.

[44] A. Tietäväinen, On pairs of Additive equations, Ann. Univ. Turku., Ser. A I 112 (1967), 1-7.

[45] A. Tietäväinen, On diagonal forms over finite fields, Ann. Univ. Turku Ser. A (1968), 1-10.

[46] R.C. Vaughan, The Hardy Littelwood method,

[47] G. Vosper, The critical pairs of subsets of a group of prime order, J. London Math. Soc. 31 (1956), 200-205.

[48] G. Vosper, Addendum to "The critical pairs of subsets of a group of prime order", J. London Math. Soc. 31 (1956), 280-282.

2

CONNECTIVITY OF CAYLEY DIGRAPHS

Ralph Tindell
EECS Department
Stevens Institute of Technology
Hoboken. NJ, U.S.A.

2.1   INTRODUCTION

The purpose of this chapter is to study several important mathematical problems related to the interconnection structure of networks. Thus the objects of our study are directed and undirected graphs. In many situations it is highly advantageous to use interconnection networks which are highly symmetric. This often simplifies computational and routing algorithms. One way in which symmetric networks may be constructed is to use interconnection schemes based on the algebraic structure called a group. Such digraphs are called Cayley digraphs and are the primary objects of study in this chapter. We shall concentrate on network connectivity, which may be viewed as measures of vulnerability of the network to global failure as the result either of failures of nodes or failures of links. Many of the results to be presented are valid for more general classes of digraphs than Cayley digraphs; where this is the case we have attempted to carry out the development at the most general level consistent with our overall goals.

2.2   TERMINOLOGY AND DEFINITIONS

A *digraph* is a pair $X = (V, E)$ where V is a finite set and E is an irreflexive relation on V. Thus E is a set of ordered pairs $(u, v) \in V \times V$ such that $u \neq v$. The elements of V are called the *vertices*, or *nodes*, of X and the elements of E are called the *arcs* of X. Arc (u, v) is said to be an *inarc* of v

41

Ding-Zhu Du and D. Frank Hsu (eds.), Combinatorial Network Theory, 41–64.
© *1996 Kluwer Academic Publishers. Printed in the Netherlands.*

and an *outarc* of u; we also say that (u,v) *originates* at u and *terminates* at v. If u is a vertex of X, then the *outdegree* of u in X is the number $d_X^+(u)$ of arcs of X originating at u and the *indegree* of u in X is the number $d_X^-(u)$ of arcs of X terminating at u. Clearly $\sum_{u \in V} d_X^+(u) = \sum_{u \in V} d_X^-(u) = |E|$. The minimum outdegree of X is $\delta^+(X) = \min\{ d_X^+(u) \mid u \in V \}$ and the minimum indegree of X is $\delta^-(X) = \min\{ d_X^-(u) \mid u \in V \}$. We denote by $\delta(X)$ the minimum of $\delta^+(X), \delta^-(X)$. A *subdigraph* of digraph $X = (V, E)$ is a digraph $X' = (V', E')$ with $V' \subset V$ and $E' \subset E$; if $V' = V$, then we say that X' is a *spanning* subdigraph of X. The *subdigraph induced* by a subset A of the vertex set V of X is the digraph $(A, \{ (u,v) \in E \mid u, v \in A \})$. In a minor abuse of notation, we shall write A for the subdigraph induced by A in places where a digraph is used in subscripts; thus the outdegree of a vertex $u \in A$ in the subdigraph induced by A is denoted by $d_A^+(u)$. The *reverse digraph* of digraph $X = (V, E)$ is the digraph $X^{(r)} = (V, \{ (v,u) \mid (u,v) \in E \})$. Digraph $X = (V, E)$ is *symmetric* if $E = E^{(r)}$ and is *antisymmetric* if $E \cap E^{(r)} = \emptyset$. An (undirected) graph is a pair $X = (V, E)$ where V is a finite set and E is a collection of two-element subsets of V. We will in general identify an undirected graph $X = (V, E)$ with the symmetric digraph $X_s = (V, E_s)$ where $E_s = \{ (u,v) \mid \{u,v\} \in E \} \cup \{ (v,u) \mid \{u,v\} \in E \}$. A digraph with exactly one vertex (and therefore no arcs) is called a *trivial* digraph. An *isomorphism* from digraph $X = (V_X, E_X)$ to digraph $Y = (V_Y, E_Y)$ is a bijection φ from V_X to V_Y such that $(u,v) \in E_X$ if and only if $(\varphi(u), \varphi(v)) \in E_Y$. Note that a digraph isomorphism φ as above induces a bijection between the arc sets of the two digraphs, also denoted by φ; thus if $x = (u,v) \in E$, then $\varphi(x) = (\varphi(u), \varphi(v))$. An isomorphism from X onto itself is called an *automorphism* of X. We denote by K_n^* the digraph with vertices the integers from 1 to n and arcs all pairs (i,j) of such integers with $i \neq j$. A digraph isomorphic to K_n^* is said to be a *complete symmetric digraph*. The *Cartesian product* of digraphs $X = (V_X, E_X)$ and $Y = (V_Y, E_Y)$ is the digraph $X \times Y$ with vertex set the Cartesian product $V_X \times V_Y$ and arc set

$$\{ ((u,a), (v,b)) \mid (u = v \text{ and } (a,b) \in E_Y) \text{ or } ((u,v) \in E_X \text{ and } a = b) \}$$

Note that in the Cartesian product above, each "vertical slice" $\{u\} \times V_Y$ induces a subdigraph isomorphic to Y and each "horizontal slice" $V_X \times \{a\}$ induces a subdigraph isomorphic to X.

A *path of length* m in X from vertex u to vertex v is a sequence $P = v_0, v_1, \ldots, v_m$ such that $u = v_0, v = v_n$ and $(v_{i-1}, v_i) \in E$ for $1 \leq i \leq n$; P is *closed* if $v_0 = v_m$, P is *simple* if $i \neq j$ implies $v_i \neq v_j$, and P is a *simple closed path* provided $v_i = v_j$ if and only if $i = j$ or $\{i,j\} = \{0, m\}$. If there exists at least one path from u to v in X we say that v is *accessible* in X from u. We say that X is *strongly connected* if for each pair u, v of vertices of X, each of u, v is accessible from the other in X. We say that X is *connected* if for any pair of vertices u, v of X, there is a sequence $u = v_0, v_1, \ldots, v_m = v$ such that for each $i, 1 \leq i \leq m$, at least one of $(u_{i-1}, u_i), (u_i, u_{i-1})$ is in E. The vertices and edges of a simple closed path in a digraph X form a subdigraph of X called a *circuit* or *directed cycle*. A subdigraph consisting of a circuit and its reversal is called a *cycle* of X. If X is strongly connected and $u, v \in V$, the *distance* in X from u to v is the length $\mathrm{dist}_X(u, v)$ of a shortest path in X from u to v. It is obvious that a shortest path between two vertices must be simple. The *diameter* of a strongly connected digraph X is $\mathrm{diam}(X) = \max\{ \mathrm{dist}_X(u, v) \mid u, v \in V \}$.

Recall that a *group* is a set G together with an associative binary operation $\cdot$ on G such that G contains an identity element 1_G for $\cdot$ and each element g of G has an inverse, which is to say an element g^{-1} such that $g \cdot g^{-1} = g^{-1} \cdot g = 1_G$. A general example of a group is the set of all permutations of a given set T, which is to say the set of all bijections from T to itself, with function composition as the group operation. This group is often called the *symmetric group* on T. An *Abelian* group is a group whose binary operation is commutative, which is to say that $g \cdot h = h \cdot g$ for all $g, h \in G$. It is usual in Abelian groups to write the operation as $+$, the inverse of element g as $-g$ and the identity as 0. The simplest finite Abelian groups are the groups $\mathbf{Z}_n$ with set $\{ 0, 1, \ldots, n-1 \}$ and with operation addition modulo n. A *homomorphism* from group G with operation $\cdot$ to group G' with operation $\star$ is a function φ from the set G to the set G' such that $\varphi(1_G) = 1_{G'}$ and $\varphi(g \cdot h) = \varphi(g) \star \varphi(h)$ for all $g, h \in G$. The set of all elements $g \in G$ with $\varphi(g) = 1_{G'}$ is called the *kernel* of the homomorphism φ. A bijective homomorphism is an *isomorphism* and an isomorphism from a group G onto itself is called an *automorphism* of G. By way of example, it should be clear that the symmetric group on a set T is uniquely determined, up to isomorphism, by the cardinality of the set T. The symmetric group on a set of cardinality n, usually taken to be the integers from 1 to n, is denoted $\mathrm{Sym}(n)$. For each element $g \in G$, the mapping ι_g given by $\iota_g(h) = g \cdot h \cdot g^{-1}$ for each $h \in G$ is an automorphism of G and is called an *inner automorphism* of G. A *normal* subset of G is a subset S invariant under all inner automorphisms, which is to say that $S = g \cdot S \cdot g^{-1}$ for all $g \in G$. It is simple to see that a set S is normal if and only if the set S^{-1} is normal, since $g \cdot S^{-1} \cdot g^{-1} = (g \cdot S \cdot g^{-1})^{-1}$. Note that in an Abelian group, the identity mapping is the only inner automorphism and thus all subsets are normal.

A *subgroup* of a group G is a subset H of G such that for any $g, h \in H$, the elements $g \cdot h$ and g^{-1} are also in H. The group operation for H is the restriction of $\cdot$ to H. For example, the kernel of any homomorphism with domain G is a subgroup of G. In fact, the kernel of an homomorphism is a *normal* subgroup, by which we mean a subgroup H which is a normal subset of G. Thus every subgroup of an Abelian group is normal. If H is an arbitrary subgroup of G and $g \in G$, then the *left coset* of H determined by g is the set $g \cdot H = \{ g \cdot h \mid h \in H \}$ and the *right coset* of H determined by g is the set $H \cdot g = \{ h \cdot g \mid h \in H \}$. The left cosets form a partition of the set G, as do the right cosets. If H is normal, then the left cosets of H form a group in which the operation $\star$ is given by $(g \cdot H) \star (h \cdot H) = (g \cdot h) \cdot H$; this group is called the *quotient* of G by H and is denoted G/H. Note that the function mapping each element g onto the left coset $g \cdot H$ is a homomorphism of G onto G/H with kernel H. In fact, if φ is a surjective homomorphism from G to G', then G' is isomorphic to the quotient of G by the kernel of φ. It is easy to see that every group G is isomorphic to a group of permutations: just assign to each $g \in G$ the "left translation" ℓ_g defined by $\ell_g(h) = g \cdot h$ for all $h \in G$. Since $\ell_g \ell_h(g') = \ell_{g \cdot h}(g')$ for any $g, h, g' \in G$, it is clear that the left translations form a subgroup Λ_G of the group of permutations of the set G and that Λ_G is isomorphic to G. Note that, except for the identity function ℓ_1, a left translation is not an automorphism of G. However, the set of all group automorphisms of G do form a subgroup of the symmetric group over G. Other interesting examples of permutation groups are the group $\text{Aut}(X)$ of all automorphisms of a digraph $X = (V, E)$ as well as the group of automorphism-induced permutations of the arc set E of X.

If S is a subset of G, then the *subgroup generated by* S, denoted by $\langle S \rangle$, is the least subgroup G containing S; $\langle S \rangle$ is in fact the intersection of all subgroups of G containing S. If $\langle S \rangle = G$, then we say that S *generates* G; if it is also the case that no proper subset of S generates G, then we say that S is a *minimal* generating set for G. Given $g \in G$ and an integer n, we define g^n inductively by setting $g^0 = 1_G$ and stipulating that $g^{k+1} = g \cdot g^k$ for all $k \geq 0$. If there is an integer n such that $g^n = 1_G$, then we define the *order* of g to be the least such integer n; if no such integer exists, we say that g has infinite order. Clearly, each element of a finite group has finite order.

We shall henceforth assume that all groups considered are finite. If G is a group and S is a subset of $G \setminus \{ 1_G \}$, where 1_G is the identity of G, we define the *Cayley digraph* $\text{Cay}(G, S)$ to be the digraph with vertices the elements of the group G and arcs all pairs of the form $(g, g \cdot s)$ with $g \in G$ and $s \in S$. The arc $(g, g \cdot s)$ is viewed as having *label* s. If the group G is Abelian, then we shall refer to $\text{Cay}(G, S)$ as an *Abelian Cayley digraph*. Every vertex of a Cayley digraph $X = \text{Cay}(G, S)$ has indegree and outdegree equal to the cardinality $|S|$ of S, and thus $\delta^+(X) = \delta^-(X) = \delta(X) = |S|$. Consistent with our identification of (undirected) graphs with symmetric digraphs, we define a *Cayley graph* to be a symmetric Cayley digraph. It should be clear that a Cayley digraph $\text{Cay}(G, S)$ is symmetric if and only if the inverse of every element of S is again in S. A path in the Cayley graph is uniquely determined by its initial vertex and the sequence of arc labels: if g is the initial vertex and $s_1, s_2, \ldots, s_m$ is the sequence of arc labels then for $1 \leq i \leq m$, the ith vertex on the path is given by $g \cdot s_1 \cdots s_i$.

Proposition 2.2.1 *Let G be a finite group and $S \subset G \setminus \{1_G\}$. Then $X = \text{Cay}(G, S)$ is strongly connected if and only if S generates G.*

Proof. It is clear from the remarks preceding the present proposition that an element of G is accessible in X from 1_G if and only if it can be written as a product of elements of S. Thus X strongly connected implies that S generates G. On the other hand, g is in the subgroup generated by S if and only if it can be written as a product of elements of $S \cup S^{-1}$. Since G is finite, each element s of S has finite order in G, and thus we may replace any occurence of s^{-1} in a product with s^{n-1}, where n is the order of s in G. Thus given elements $h, g \in G$, we may write $h^{-1} \cdot g$ as a product of elements of S and this product determines a path in X from h to g. □

Let Φ be a subgroup of the symmetric group over a set U. We say that Φ *acts transitively* on a subset T of U if for any $s, t \in T$, there exists a permutation $\varphi \in \Phi$ with $\varphi(s) = t$. If Φ acts transitively on the entire set U, then we say that Φ is a *transitive group of permutations* on U. By way of example, we note that group Λ_G of left translations of a group G acts transitively on the whole group G, since given any $h_1, h_2 \in G$ we have $\ell_{h_2 h_1^{-1}}(h_1) = h_2$. Moreover, since $\ell_g(h \cdot s) = \ell_g(h) \cdot s$ for any $g, h, s \in G$, Λ_G is a subgroup of $\text{Aut}(\text{Cay}(G, S))$ for any subset S of $G \setminus \{1\}$. In fact, the left translations are exactly the automorphisms of $\text{Cay}(G, S)$ which preserve the arc labels. What we have shown is that every Cayley graph is a *vertex-transitive* digraph, by which we mean a digraph X whose automorphism group $\text{Aut}(X)$ acts transitively on the vertex set of X. Note that every vertex of a vertex-transitive digraph X has indegree and outdegree $\delta^+(X) = \delta^-(X) = \delta(X)$. If $X = (V, E)$ is either a digraph or a graph, then the automorphism induced permutations of E form a permutation group over E. If this group acts transitively on E, then we say that X is *arc-transitive* if X is a digraph and that X is *edge-transitive* if X is a graph. A graph is said to be *1-transitive* if it is arc-transitive when viewed as a symmetric digraph. It is easy to see that the complete bipartite graph $K_{1,2}$ with parts of size 1 and 2, respectively, is edge-transitive but not 1-transitive, so the two concepts are not equivalent. It is however interesting to note that an Abelian Cayley graph is edge-transitive if and only if it is 1-

transitive. This follows from the fact that the function φ taking each group element g to its inverse $-g$ is a digraph automorphism and thus for each $s \in S$, the automorphism $\ell_{-s}\varphi$ maps the arc $(0, s)$ to its reversal $(s, 0)$.

We have noted above that left translations of a group G are digraph isomorphisms of $\mathrm{Cay}(G, S)$ for any $S \subset G \setminus \{1_G\}$. This is due to the fact that the out arcs of a vertex g are determined by right multiplication by elements of S. The *right translation* by $g \in G$ is the permutation r_g of G given by $r_g(h) = h \cdot g$ for each $h \in G$. Right translations of G are not digraph automorphisms of $\mathrm{Cay}(G, S)$ for all subsets S of $G \setminus \{1_G\}$. However, we may characterize the Cayley digraphs for which right translations are digraph isomorphisms.

Proposition 2.2.2 *Let G be a finite group and let S be a subset of $G \setminus \{1_G\}$. Then the right translations of G are digraph automorphisms of $X = \mathrm{Cay}(G, S)$ if and only if S is a normal subset of G.*

Proof. For each $h \in G$, the right translation r_h is a digraph automorphism of X if and only if $r_h(1_G, s) = (h, s \cdot h)$ is an arc of X for any $s \in S$. But this is equivalent to saying that $s \cdot h = h \cdot s'$ for some $s' \in S$, which holds if and only if $h^{-1} \cdot s \cdot h \in S$. Thus all right translations of G are automorphisms of X if and only if $h^{-1} \cdot S \cdot h = S$ for all $h \in G$, which is to say that S is normal. $\square$

2.3   EDGE-CONNECTIVITY

Let $X = (V, E)$ be a digraph. An *arc disconnecting set* of X is a subset W of E such that $X \setminus W = (V, E \setminus W)$ is not strongly connected. An arc disconnecting set is *minimal* if no proper subset of W is an arc disconnecting set of X and is a *minimum arc disconnecting set* if no other arc disconnecting set has smaller cardinality than W. The *arc connectivity* $\lambda(X)$ of a nontrivial digraph X is the cardinality of a minimum arc disconnecting set of X. The *positive arc neighborhood* of a subset A of V is the set $\omega_X^+(A)$ of all arcs which initiate at a vertex of A and terminate at a vertex of $V \setminus A$. The *negative neighborhood* of subset A of V is the set $\omega_X^-(A)$ of arcs which initiate in $V \setminus A$ and terminate in A. Thus $\omega_X^-(A) = \omega_X^+(V \setminus A)$. Arc neighborhoods of proper, nonempty subsets of V, often called *cuts*, are clearly arc disconnecting sets. Thus for any proper, nonempty subset A of V, $|\omega_X^+(A)| \geq \lambda(X)$. If we consider the cases where A consists of a single vertex or the complement of a single vertex, we easily see that $\delta(X) \geq \lambda(X)$. An equivalent, and for our purposes more convenient, definition of arc connectivity is provided by the following proposition.

Proposition 2.3.1 *If W is an arc disconnecting set of a strongly connected, nontrivial digraph $X = (V, E)$, then there is a proper, nonempty subset A of V such that $\omega_X^+(A) \subset W$. Thus*

$$\lambda(X) = \min\{\,|\omega_X^+(A)|\mid A \text{ is a proper, nonempty subset of } V\,\}.$$

Proof. Choose $u, v \in V$ such that v is not accessible in $X \setminus W$ from u and define $A = \{\,w \in V \mid w \text{ is accessible in } X \setminus W \text{ from } u\,\}$. Since $u \in A$ and $v \in V \setminus A$, A is a proper, nonempty subset of V. It is clear that every arc $(a, b) \in \omega^+(A)$ must be in W since a is accessible in $X \setminus W$ from u and b is not. □

Let $X = (V, E)$ be a nontrivial digraph. If X is either symmetric (i.e., a graph) or vertex-transitive, then $d_X^+(u) = d_X^-(u)$ for every vertex u of X. Any digraph with this property is said to be a *balanced* digraph. The balanced property may be restated as $|\omega_X^+(\{\,u\,\})| = |\omega_X^-(\{\,u\,\})|$. It is simple to prove by induction on the cardinality of A that this property holds for balanced digraphs when $\{\,u\,\}$ is replaced by an arbitrary proper, nonempty subset A of V. Since the class of balanced digraphs contains the classes of digraphs of interest to us in this section, we will develop our basic results in this setting.

Let $X = (V, E)$ be a balanced digraph. An *arc fragment* of X is a proper, nonempty subset of V whose positive arc neighborhood has cardinality $\lambda(X)$. As noted earlier, we then have $|\omega_X^+(V \setminus A)| = |\omega_X^-(A)| = \lambda(X)$ as well. Therefore A is an arc fragment of X if and only if $V \setminus A$ is an arc fragment of X. From the point of view of network design, it is obviously desirable that the arc connectivity of the network be as large as possible, and thus one is interested in digraphs which realize the upper bound of $\delta(X)$. One of the things we shall do in this section is to prove that every strongly connected, vertex-transitive digraph has this property. This result was originally proved by Mader [12] in the undirected case and later generalized to digraphs by Hamidoune [6]. In addition to having $\lambda = \delta$, one would also like to minimize the number of minimum cuts in the network. Clearly, the set of inarcs of a vertex of indegree δ and the set of outarcs of a vertex of outdegree δ will be minimum cuts, so the number of such vertices gives a lower bound on the number of minimum cuts. We shall be interested in digraphs which realize this lower bound. Thus we define a digraph X to be *super arc-connected*, or more simply, *super-λ*, if every minimum cut of X is either the set of inarcs of some vertex or the set of outarcs of some vertex. Notice that a digraph X with $\lambda(X) < \delta(X)$ is by definition not super arc-connected. The super-λ property was originally introduced for undirected graphs by Bauer, Boesch, Suffel and Tindell [1].

We now proceed to develop tools that will enable us to prove the basic results on the arc-connectivity properties of transitive digraphs. A basic first step is the following proposition.

Proposition 2.3.2 ([10]) *Let $X = (V, E)$ be a strongly connected, balanced digraph and let A and B be arc fragments of X such that $A \not\subset B$ and $B \not\subset A$. If $A \cap B \neq \emptyset$ and $A \cup B \neq V$, then each of the sets $A \cap B, A \cup B, A \setminus B$ and $B \setminus A$ is an arc fragment of X.*

Proof. It will be convenient to introduce the following notation: given subsets U, U' of V, $E[U, U'] = \{ (u, u') \in E \mid u \in U \text{ and } u' \in U' \}$. We may then partition the arc neighborhoods of the sets under consideration as follows:

$$
\begin{aligned}
\omega^+(A \cap B) &= E[A \cap B, V \setminus (A \cup B)] \cup E[A \cap B, A \setminus B] \cup E[A \cap B, B \setminus A] \\
\omega^+(A \cup B) &= E[A \cap B, V \setminus (A \cup B)] \cup E[A \setminus B, V \setminus (A \cup B)] \\
&\quad \cup E[B \setminus A, V \setminus (A \cup B)] \\
\omega^+(A) &= E[A \cap B, V \setminus (A \cup B)] \cup E[A \setminus B, V \setminus (A \cup B)] \\
&\quad \cup E[A \cap B, B \setminus A] \cup E[A \setminus B, B \setminus A] \\
\omega^+(B) &= E[A \cap B, V \setminus (A \cup B)] \cup E[B \setminus A, V \setminus (A \cup B)] \\
&\quad \cup E[A \cap B, A \setminus B] \cup E[B \setminus A, A \setminus B]
\end{aligned}
$$

It is thus evident that

$$
|\omega^+(A \cap B)| + |\omega^-(V \setminus (A \cup B))| \leq |\omega^+(A)| + |\omega^+(B)| = 2\lambda(X).
$$

Since each of $|\omega^+(A \cap B)|, |\omega^-(V \setminus (A \cup B))|$ is at least $\lambda(X)$, we can conclude that $A \cap B$ and $A \cup B$ are arc fragments. It then follows that $V \setminus (A \cup B)$ is an arc fragment and hence that $B \setminus A = B \cap (V \setminus (A \cup B))$ and $A \setminus B = A \cap (V \setminus (A \cup B))$ are also arc fragments. □

Let $X = (V, E)$ be a strongly connected balanced digraph. An arc fragment of least possible cardinality is called a λ-*atom* of X and a nontrivial arc fragment of least possible cardinality is called a λ-*superatom* of X. The idea of atoms first arose in connection with vertex-connectivity in the work of Mader, Watkins, and Hamidoune, which we shall consider in the next sections. The use of λ-atoms is due to Mader [12] and Hamidoune [6]; the use of λ-superatoms is due to Tindell [15], and fully exploited in the paper [10] of Hamidoune and Tindell. We shall not, in fact, have occasion to use λ-atoms here as we can derive the desired results using λ-superatoms.

A desirable property one wishes any type of atom to have is that, if nontrivial, they form "imprimitive blocks" for the automorphism group of the digraph. To be precise, an *imprimitive block* for a group Φ of permutations of a set T is a proper, nontrivial subset A of T such that if $\varphi \in \Phi$ then either $\varphi(A) = A$ or $\varphi(A) \cap A = \emptyset$. The following proposition indicates why imprimitivity is so useful. Recall that a set of vertices of a digraph is *independent* if its induced subdigraph contains no arcs.

Proposition 2.3.3 *Let $X = (V, E)$ be a graph or digraph and let Y be the subdigraph induced by an imprimitive block A of X. Then*

1. *If X is vertex-transitive then so is Y;*

2. *If X is a strongly connected arc-transitive digraph or a connected edge-transitive graph and A is a proper subset of V, then A is an independent subset of X;*

3. *If $X = Cay(G, S)$ and A contains the identity of G, then A is a subgroup of G.*

Proof. To prove conclusion (1), we merely note that given $u, v \in A$, there is an automorphism φ of X with $\varphi(u) = \varphi(v)$; since $\varphi(A) \cap A \neq \emptyset$, $\varphi(A) = A$ and thus φ induces an automorphism of the induced subdigraph Y. As for (2), suppose there is an arc (or edge) y with both endpoints in A. Since X is strongly connected, $\omega_X^+(A) \neq \emptyset$ and thus there must be an automorphism φ of X such that $\varphi(y) \in \omega_X^+(A)$. But then we have $\varphi(A) \cap A \neq \emptyset$ and $\varphi(A) \cap (V \setminus A) \neq \emptyset$, which is impossible. Therefore A must be an independent set of vertices and (2) is proved. Finally, for (3), consider the case where A is an imprimitive block of $Cay(G, S)$ containing the identity 1 of G. Then for any $g \in A$, $1 \in \ell_{g^{-1}}(A) = g^{-1} \cdot A$ and hence $g^{-1} \cdot A = A$. Thus $g, h \in A$ implies that $g^{-1} \cdot h \in A$, from which it is obvious that A is a subgroup of G. $\qquad\square$

If X is a strongly connected balanced digraph which is not super-λ, then clearly any automorphic image of a λ-superatom of X is again a λ-superatom of X. Thus to prove that λ-superatoms are imprimitive blocks we need only show that distinct λ-superatoms are disjoint, which do next.

Proposition 2.3.4 *Let $X = (V, E)$ be a strongly connected balanced digraph which is not a symmmetric cycle, is not super arc-connected and has $\delta(X) \geq 2$. If $\delta(X) > 2$ or X is vertex-transitive, then distinct λ-superatoms of X are disjoint. Thus λ-superatoms are imprimitive blocks.*

Proof. Suppose to the contrary that that there are distinct λ-superatoms A, B of X with $A \cap B \neq \emptyset$. By Lemma 2.3.2, each of $A \cap B, A \setminus B$ and $B \setminus A$ is a λ-fragment which is a proper subset of a λ-superatom. Therefore, each of these sets must have cardinality 1 so that we may assume $A = \{v, u\}$ and $B = \{u, w\}$ with $v \neq w$. We thus have $d_A^+(v) = d_A^-(u) \leq 1$ and $d_A^+(u) = d_A^-(v) \leq 1$. Therefore

$$
\begin{aligned}
\delta(X) &\geq \lambda(X) \\
&= |\omega^+(A)| \\
&= d_X^+(v) - d_A^+(v) + d_X^+(u) - d_A^+(u) \\
&\geq 2\delta(X) - 2,
\end{aligned}
$$

so that $\delta(X) \leq 2$ and hence $\delta(X) = 2$. Note that since this is impossible if $\delta(X) > 2$, the conclusion holds in this case. Thus we assume that X is vertex-transitive. Since $\delta(X) = 2$, we conclude that $d_A^+(v) = d_A^+(u) = 1$. Applying the preceding argument with B in place of A, we obtain $d_B^+(w) = d_B^+(u) = 1$. It then follows that the arcs incident with u are $(u, v), (v, u), (u, w)$ and (w, u). Since X is vertex-transitive, we see that it is a symmetric digraph of degree 2 that is strongly connected and hence is a symmetric cycle. This contradiction completes the proof.

Proposition 2.3.5 *Let $X = (V, E)$ be a strongly connected balanced digraph such that $\delta(X) \geq 2$ and X is not super-λ. Then (1) every λ-superatom has cardinality at most $\lfloor |V|/2 \rfloor$; and (2) if X is vertex-transitive, then each λ-superatom induces a strongly connected subdigraph of X.*

Proof. Let A be a λ-superatom of X. Then $V \setminus A$ is also a nontrivial arc fragment of X, so $|A| \leq |V \setminus A|$ and thus $|A| \leq \lfloor |V|/2 \rfloor$. Next suppose that X is vertex-transitive. Then, by Proposition 2.3.4, A is an imprimitive block of X and hence, by Proposition 2.3.3, the subdigraph Y induced by A is vertex-transitive. Clearly Y must be connected as otherwise we could find a cut with fewer than $\lambda(X)$ arcs. It then follows that Y is strongly connected since is easy to see that a connected vertex-transitive digraph must be stongly connected. $\square$

Corollary 2.3.6 *Let X be a strongly connected digraph which is not a symmetric cycle and has $\delta(X) \geq 2$. If X is either an arc-transitive digraph or an edge transitive graph, then X is super-λ.*

Proof. If X satisfies the hypothesis of Proposition 2.3.4, then it contains a λ-superatom A, which is an imprimitive block of X. But by Proposition 2.3.5, the subdigraph Y induced by A is strongly connected. Since $|A| \geq 2$, Y contains at least one arc, which contradicts part 2 of Proposition 2.3.3. Thus we may assume X does not satisfy the hypothesis of Proposition 2.3.4. Since arc-transitive digraphs are vertex-transitive, we thus must have that X is an edge transitive graph which is not a cycle and satisfies $\delta(X) \leq 2$. It is not hard to see that the only such graphs are the bipartite graphs with at least one part of size $\delta(X)$, where $\delta(X)$ is either 1 or 2. Since each such graph is super-λ, the proof is complete.

A *clique* in a digraph X is a complete symmetric subdigraph of X. If A is a subset of the vertex set of a digraph X, the value of δ on the subdigraph induced by A will, in a minor abuse of notation, be denoted by $\delta(A)$.

Proposition 2.3.7 *Let $X = (V, E)$ be a strongly connected vertex-transitive digraph which is not a symmetric cycle, is not super arc-connected and has $\delta(X) \geq 2$. Then*

1. $\lambda(X) = \delta(X)$

2. *$A \subset V$ is a λ-superatom of X if and only if A induces a clique of X and $|A| = \delta(X)$.*

Proof. Since cliques are super arc-connected, we may assume that $2 \leq \delta(X) \leq n - 2$. First suppose A is a λ-superatom of X. Then $|A|(\delta(X) - \delta(A)) = \lambda(X) \leq \delta(X) = \delta(A) + (\delta(X) - \delta(A))$. Thus $(|A| - 1)(\delta(X) - \delta(A)) \leq \delta(A) \leq |A| - 1$. But this has several consequences: (1) equality holds in the first inequality, so that $\lambda(X) = \delta(X)$; (2) $\delta(A) = |A| - 1$, so that A is a clique of X; and (3) $\delta(X) - \delta(A) = 1$, so that $\delta(X) = |A|$. All that remains is to show that every subset A of V of size $\delta(X)$ that induces a clique is a λ-superatom of X. But each vertex in such a subset would have exactly one outarc in X terminating at a vertex in $V \setminus A$ and thus $\omega^+(A) = |A| = \delta(X) = \lambda(X)$, so that A is a λ-fragment of X. Since we showed in the first part of this proof that the λ-superatoms of X have cardinality $\delta(X)$, it follows that A is a λ-superatom of X. □

Since a digraph X which is super-λ satisfies $\lambda(X) = \delta(X)$, we have as a corollary the maximal connectivity result of Mader [12] (undirected graphs) and Hamidoune [6] (digraphs).

Corollary 2.3.8 *Every strongly connected, vertex-transitive digraph X satisfies $\lambda(X) = \delta(X)$.*

The remaining results of this section are from the paper [10] by the present author and Hamidoune.

Theorem 2.3.9 (Local Characterization) *Let $X = (V, E)$ be a strongly connected vertex-transitive digraph which is neither a clique nor a symmetric cycle and has $\delta(X) \geq 2$. Then X is not super arc-connected if and only if X contains a clique of size $\delta(X)$.*

Proof. The only if part of this theorem is immediate from Proposition 2.3.7. Thus suppose that X contains a clique with vertex set A and that $|A| = \delta(X)$. Since X is not a clique, $1 < \delta(X) = |A| < n - 1$. Moreover, $|\omega^+(A)| = |A| = \delta(X) = \lambda(X)$. It then follows that A is a nontrivial fragment of X with $|V \setminus A| > 1$, and hence that X is not super arc-connected.

Theorem 2.3.10 (Algebraic Characterization) *Let $Cay(G, S)$ be a strongly connected Cayley digraph with $|S| \geq 2$. If $Cay(G, S)$ is not a symmetric cycle, then $Cay(G, S)$ is not super arc-connected if and only if $S = (H \setminus \{1\}) \cup \{t\}$ for some nontrivial subgroup H of G and some element t of $G \setminus H$.*

Proof. To prove the if part of the theorem, suppose $S = (H \setminus \{1\}) \cup \{t\}$ for some nontrivial subgroup H of G and some element t of $G \setminus H$. Then $\omega^+(H) = (H \cdot S) \setminus H = (H \cdot (H \setminus \{1\}) \cup H \cdot t) \setminus H = (H \cup H \cdot t) \setminus H = H \cdot t$ and hence $|\omega^+(H)| = |H| = |S| = \lambda(Cay(G, S))$. It then follows that $Cay(G, S)$ is not super arc-connected.

To prove the only if part of the theorem, suppose $Cay(G, S)$ is not super arc-connected and let H be the λ-superatom containing the identity of G. By Propositions 2.3.3 and 2.3.4, H is a subgroup of G. By the Local Characterization Theorem, H induces a complete symmetric subdigraph of cardinality $\delta(Cay(G, S)) = |S|$ and hence each element has exactly one outarc which terminates outside H. It then follows that $|S \cap H| = |S| - 1$ and thus, since $1 \notin S$, we have $S \cap H = H \setminus \{1\}$. Since $|H \setminus \{1\}| = |S| - 1$, we see that $S = (H \setminus \{1\}) \cup \{t\}$ for some $t \in G \setminus H$ and thus we are done.

Corollary 2.3.11 *A connected Abelian Cayley graph is not super-λ if and only if it is either a cycle or is isomorphic to $K_m^* \times K_2^*$ for some $m \geq 2$.*

Proof.   Let Cay(G, S) be a connected symmetric Cayley digraph with G Abelian and suppose that Cay(G, S) is not super edge-connected and is not a cycle. Then $S = -S$ and, by the Algebraic Characterization Theorem, $S = (H \setminus \{0\}) \cup \{t\}$ for some nontrivial subgroup H of G and some $t \in G \setminus H$. It then follows that $t = -t$ and hence that $2t = 0$. Therefore, H has exactly two distinct left cosets, namely H and $t + H$, and each of these cosets has cardinality $m = |H| \geq 2$. Thus the Cayley graph consists of the two cliques induced by H and $H + t$, which have size m, together with a perfect matching between the cliques, from which it is obvious that Cay(G, S) is isomorphic to $K_m^* \times K_2^*$. The if part of the theorem may be established by verifying that $K_m^* \times K_2^*$ is isomorphic to Cay($\mathbf{Z}_m \times \mathbf{Z}_2, (\mathbf{Z}_m \times \{0\} \setminus \{(0,0)\}) \cup \{(0,1)\}$). $\square$

A Cayley graph over $\mathbf{Z}_n$, the integers modulo n, is called a *circulant*. We may now derive a result first proved by Boesch and Wang [2].

Corollary 2.3.12 *A connected circulant is not super-λ if and only if it is a cycle of length 4 or more or is isomorphic to $K_m^* \times K_2^*$ for some odd integer $m > 1$.*

Proof.   We need only verify that $K_m^* \times K_2^*$ is a circulant if and only if m is odd. It is easy to see that if m is odd, then Cay($\mathbf{Z}_{2m}, \{m\} \cup \{2i \mid 1 \leq i \leq m - 1\}$) is isomorphic to $K_m^* \times K_2^*$. To prove the converse, suppose that $K_m^* \times K_2^*$ is isomorphic to Cay($\mathbf{Z}_{2m}, S$) for some generating set S of $\mathbf{Z}_{2m}$. Then by the Algebraic Characterization Theorem, $S = (H \setminus \{1\}) \cup \{t\}$ for some nontrivial subgroup H of $\mathbf{Z}_{2m}$ and element t of $\mathbf{Z}_{2m} \setminus H$. Since the Cayley digraph must be symmetric, $-t \in S = (H \setminus \{1\}) \cup \{t\}$. Since H is a subgroup not containing t, it cannot contain $-t$ and thus $-t = t$ in $\mathbf{Z}_{2m}$, so that $t = m$. Moreover, H has exactly two distinct cosets in $\mathbf{Z}_{2m}$ and the only such subgroup is the set of even elements. Since $t = m \notin H$, we may conclude that m is odd. $\square$

We conclude the present section with the statement, without proof, of the characterization theorem for Abelian Cayley digraphs which are not super-λ. The proof may be found in [10].

Theorem 2.3.13 *The only Abelian Cayley digraphs that are not super-λ are the symmetric cycles and the Cayley digraphs isomorphic to*

$$Cay(\mathbf{Z}_{rp} \times \mathbf{Z}_m, \{(ip, j) \mid 0 \leq i < r, 0 \leq j < m\} \setminus \{(0,0)\}),$$

for some positive integers m, r, p with $p \geq 2$ and $mr \geq 2$.

2.4 CONNECTIVITY AND ATOMS

Let $X = (V, E)$ be a digraph. A *vertex disconnecting set* of X is a subset U of V such that the subdigraph $X \setminus U$ induced by $V \setminus U$ is either trivial or is not strongly connected. A vertex disconnecting set is *minimal* if no proper subset of U is a vertex disconnecting set of X and is a *minimum vertex disconnecting set* if no other vertex disconnecting set has smaller cardinality than U. The *connectivity* $\kappa(X)$ of a nontrivial digraph X is the cardinality of a minimum vertex disconnecting set of X. In this section we shall develop some basic concepts and tools for studying connectivity and apply them to the study of transitive graphs and digraphs, especially Cayley graphs and digraphs.

The *positive neighborhood* of a subset A of V is the set $N_X^+(A)$ of all vertices of $V \setminus A$ which are targets of arcs initiating at a vertex of A. The *positive closure* $\mathbf{C}_X^+(A)$ of A is the union of A and $N_X^+(A)$. The *negative neighborhood* of subset A of V is the set $N_X^-(A)$ of vertices of $V \setminus A$ which are the initial vertices of arcs which terminate at a vertex of A. The *negative closure* $\mathbf{C}_X^-(A)$ of A is the union of A and $N_X^-(A)$. Note that the negative neighborhood of A in X is the same set as the positive neighborhood of A in the reverse digraph $X^{(r)}$ of X. It is trivial to verify from definitions that $\mathbf{C}_X^+(A \cup B) = \mathbf{C}_X^+(A) \cup \mathbf{C}_X^+(B)$ and $\mathbf{C}^+(A \cap B) \subset \mathbf{C}^+(A) \cap \mathbf{C}^+(B)$ for any subsets A, B of V.

If A is a nonempty subset of V with $\mathbf{C}_X^+(A) \neq V$, then the positive neighborhood of A is clearly a vertex disconnecting set for X. Thus for each such set A, $|N_X^+(A)| \geq \kappa(X)$. If we consider the cases where A consists of a single vertex or the complement of a single vertex, we easily see that $\delta(X) \geq \kappa(X)$. An equivalent, and for our purposes more convenient, definition of vertex connectivity is provided by the following proposition.

Proposition 2.4.1 *Let $X = (V, E)$ be a strongly connected, nontrivial digraph which is not a complete symmetric digraph. If U is a vertex disconnecting set of X, then there is a proper, nonempty subset A of V such that $N_X^+(A) \subset U$ and $\mathbf{C}_X^+(A) \neq V$. Thus*

$$\kappa(X) = \min\{\, |N_X^+(A)| \mid A \text{ is a nonempty subset of } V \text{ with } \mathbf{C}_X^+(A) \neq V \,\}.$$

Proof. Choose $u, v \in V$ such that v is not accessible in $X \setminus U$ from u and define $A = \{\, w \in V \mid w \text{ is accessible in } X \setminus U \text{ from } u \,\}$. Since $u \in A$, A is nonempty. If $w \in N_X^+(A)$, there exists a vertex $w' \in A$ such that (w', w) is an arc of X. Thus if w were not in U it would be accessible from u in $X \setminus U$ and hence would be in A, which is impossible. Thus $N_X^+(A) \subset U$. Since v is not in

U, it is not in $N_X^+(A)$; since it is not accessible in $X \setminus U$ from u, it cannot be in A. Therefore $\mathbf{C}_X^+(A) \neq V$, and the proof is complete. $\qquad \square$

Let $X = (V, E)$ be a strongly connected, nontrivial digraph which is not a complete symmetric digraph. A subset F of V is called a *positive (respectively, negative) fragment* of X if $|N_X^+(F)| = \kappa(X)$ and $\mathbf{C}_X^+(F) \neq V$ (respectively, $|N_X^-(F)| = \kappa(X)$ and $\mathbf{C}_X^-(F) \neq V$). A fragment of minimum cardinality is called an *atom* of X. Note that an atom may be a either a positive fragment or a negative fragment or both. An atom which is a positive fragment is called a *positive atom* and an atom which is a negative fragment is called a *negative atom*. The notion of atom was introduced by Watkins [16] for undirected graphs and extended to digraphs by Chaty [3].

Proposition 2.4.2 *Let $X = (V, E)$ be a nontrivial, strongly connected digraph which is not a complete symmetric digraph. Then (1) $\kappa(X) = \delta(X)$ if and only if every atom of X has cardinality 1; and (2) if $\kappa(X) < \delta(X)$, then each atom has cardinality at most $\lfloor (|V| - \kappa(X))/2 \rfloor$ and induces a strongly connected subdigraph of X.*

Proof. The first conclusion is obvious. We prove the remaining conclusions for positive atoms; the proof for negative atoms is analogous. Let A be a positive atom of X. Since $V \setminus \mathbf{C}_X^+(A)$ is a negative fragment of X it must have cardinality greater than or equal to $|A|$. Since $V = A \cup N_X^+(A) \cup (V \setminus \mathbf{C}_X^+(A))$ and $|N_X^+(A)| = \kappa(X)$, conclusion (2) follows. To see (3), let Y be the subdigraph induced by A. Then Y must have a strong component F such that every arc of Y initiating in F terminates in F. But then $N_X^+(V(F)) \subset N_X^+(A)$ and hence $V(F)$ is a positive fragment of X contained in A. Since A is an atom, it then follows that $F = Y$ and we have shown that Y is strongly connected. $\qquad \square$

As in the case with λ-atoms, we wish to prove that atoms of X are imprimitive blocks of X. This fundamental fact is a corollary of the following result, due to Mader [13] in the undirected case and Hamidoune [5] in the directed case.

Proposition 2.4.3 *Let $X = (V, E)$ be a strongly connected digraph which is not a complete symmetric digraph and let A be a positive (respectively, negative) atom of X. If B is a positive (respectively, negative) fragment of X with $A \cap B \neq \emptyset$, then $A \subset B$.*

Proof. We may assume that A and B are positive fragments, since the result for the negative case will then follow by considering the reverse digraph $X^{(r)}$.

If $|N^+(A \cap B)| = \kappa(X)$ then $A \cap B$ is a positive fragment contained in the atom A so that $A \cap B = A$ and we are done. Thus we assume $|N^+(A \cap B)| > \kappa(X)$ and derive a contradiction. Since the negative fragment $V \setminus \mathbf{C}^+(B)$ has cardinality at least that of A and $|V| = |V \setminus \mathbf{C}^+(B)| + |\mathbf{C}_X^+(B)|$, we see that $|V| \geq |A| + |\mathbf{C}_X^+(B)| = |A| + |B| + \kappa(X) = |A \cup B| + |A \cap B| + \kappa(X)$. We then have the following.

$$
\begin{aligned}
&\quad |A \cup B| + |N^+(A \cup B)| \\
&= \; |\mathbf{C}^+(A \cup B)| \\
&= \; |\mathbf{C}^+(A) \cup \mathbf{C}^+(B)| \\
&= \; |\mathbf{C}^+(A)| + |\mathbf{C}^+(B)| - |\mathbf{C}^+(A) \cap \mathbf{C}^+(B)| \\
&\leq \; |\mathbf{C}^+(A)| + |\mathbf{C}^+(B)| - |\mathbf{C}^+(A \cap B)| \\
&= \; |A| + |N^+(A)| + |B| + |N^+(B)| - (|A \cap B| + |N^+(A \cap B)|) \\
&= \; (|A| + |B| - |A \cap B|) + (2\kappa(X) - |N^+(A \cap B)|) \\
&< \; |A \cup B| + \kappa(X).
\end{aligned}
$$

Since $|V| > |A \cup B| + \kappa(X)$, we see that $\mathbf{C}^+(A \cup B) = (A \cup B) \cup N^+(A \cup B) \neq V$, so that $N^+(A \cup B)$ is a vertex cut with cardinality less than $\kappa(X)$, a contradiction that completes the proof. $\qquad \square$

Corollary 2.4.4 *If $X = (V, E)$ is a strongly connected digraph which is not a complete symmetric digraph, then distinct positive (respectively, negative) atoms of X are disjoint. Thus if $\kappa(X) < \delta(X)$, the atoms of X are imprimitive blocks of X.*

Corollary 2.4.5 *Let $X = Cay(G, S)$ where $S \subset G \setminus \{1_G\}$ generates G. Then the atom A of X containing 1_G is the subgroup of G generated by $S \cap A$.*

Proof. We know from Proposition 2.3.3 that A is a subgroup of G. Also, $N^+(A) = (A \cdot S) \setminus A = A \cdot (S \setminus A)$. If A' is the subgroup of G generated by $S \cap A$, then A' is a subgroup of A. Moreover, $S \setminus A' = S \setminus A$ so that $N^+(A') = A' \cdot (S \setminus A') \subset A \cdot (S \setminus A) = N^+(A)$. Therefore A' is a positive fragment contained in the atom A and hence equals A. $\qquad \square$

We now have the following analogue of Corollary 2.3.6, with essentially the same proof, which we omit.

Corollary 2.4.6 *Let X be a strongly connected digraph which is not a symmetric cycle. If X is either an arc-transitive digraph or an edge transitive graph, then $\kappa(X) = \delta(X)$.*

Corollary 2.4.7 *Let $X = (V, E)$ be a strongly connected digraph which is not a complete symmetric digraph and let A be an atom of X. Then $\delta(X) \leq \delta(A) + \kappa(X)$.*

Proof. We assume that A is a positive atom; the proof when A is a negative atom is similar. Let $v \in A$ be a vertex of outdegree $\delta(A)$ in the subdigraph induced by A. Then $N_X^+(v) \setminus N_A^+(v) \subset N_X^+(A)$ and thus $\delta(X) \leq d_X^+(v) = |N_X^+(v)| \leq |N_A^+(v)| + |N_X^+(A)| = \delta(A) + \kappa(X)$. $\square$

The next result was proved in the undirected case (i.e., $S' = S^{-1}$) by Godsil [4] and later for the directed case by Hamidoune [7].

Theorem 2.4.8 *If S is a minimal generating set for finite group G and $S' \subset S^{-1}$, then $X = Cay(G, S \cup S')$ satisfies $\kappa(X) = \delta(X)$.*

Proof. We suppose to the contrary that $\kappa(X) < \delta(X)$ and deduce a contradiction. If we let H be the atom of X containing the identity of G, then H is the (nontrivial) subgroup of X generated by $(S \cup S') \cap H$. Moreover, $N^+(H) = H \cdot ((S \cup S') \setminus H)$. Now H is a subgroup and $S' \subset S^{-1}$, so that $S' \cap H = S' \cap (S \cap H)^{-1}$ and hence H is generated by $S \cap H$. Note that $|H| > |H \cap (S \cup S')|$ since $1_G \notin S$. If all the cosets of H by elements of $(S \cup S') \setminus H$ were distinct then we would have $|N^+(H)| = |H \cdot ((S \cup S') \setminus H)| = |H||(S \cup S') \setminus H| \geq |H| + |(S \cup S') \setminus H| > |S \cup S'|$, which is impossible. Thus there exist distinct elements $s, t \in (S \cup S')$ such that $H \cdot t = H \cdot s$. If either s and t are both elements of S, or $s \in S$ and $t \in (S \setminus \{s\})^{-1}$, then G is generated by $S \setminus \{t\}$ which contradicts the minimality of S. Thus we may conclude that there is an element $s \in S$ such that $s^{-1} \in S'$, $s \neq s^{-1}$ and $H \cdot s = H \cdot s^{-1}$. It then follows that $s^2 \in H$ and thus, since S is minimal, that $s^2 \in H \setminus (S \cup S')$. Since $s \neq s^{-1}$, $s^2 \neq 1_G$. Therefore H contains the disjoint union of the sets $\{1_G\}, \{s^2\}$, and $(S \cup S') \cap H$ and hence $|H| \geq |(S \cup S') \cap H| + 2$. Note that $|(S \cup S') \setminus H| \leq 2|S \setminus H|$. Therefore we may deduce that $(|H| - 2) + 2|S \setminus H| \geq |(S \cup S') \cap H| + |(S \cup S') \setminus H| = |S \cup S'| = \delta(X) > \kappa(X) = |N_X^+(H)| = |H \cdot ((S \cup S') \setminus H)| \geq |H \cdot (S \setminus H)| = |H||S \setminus H|$. But then we have $|H| - 2 > (|H| - 2)|S \setminus H|$, which is impossible since $|H| \geq 2$ and $|H \setminus S| \geq 1$. $\square$

Proposition 2.4.9 (Hamidoune[5]) *If $X = (V, E)$ is a strongly connected vertex-transitive digraph, then the cardinality of an atom of X is at most $\kappa(X)$.*

Proof.   Let A be an atom, which we suppose to be a positive fragment. The proof in the case that A is a negative fragment is analogous. First we claim that for any $v \in N_X^+(A)$, $|N_X^-(v) \cap A| \leq \delta(X) - \delta(A)$. To see this, let B be the positive atom containing v. Note that the subgraphs induced by B and A are isomorphic, and thus $\delta(A) = \delta(B)$. Moreover A and B are disjoint, so that $N_X^-(v) \cap A \subset N_X^-(v) \cap (V \setminus B)$ and hence $|N_X^-(v) \cap A| \leq \delta(X) - \delta(B) = \delta(X) - \delta(A)$, and the claim is thus proved. To complete the proof of the lemma, we note that

$$
\begin{aligned}
|A|(\delta(X) - \delta(A)) &= |\omega^+(A)| \\
&= \sum_{v \in N_X^+(A)} |N_X^-(v) \cap A| \\
&\leq |N_X^+(A)|(\delta(X) - \delta(A)) \\
&= \kappa(X)(\delta(X) - \delta(A))
\end{aligned}
$$

and the result follows immediately.

Theorem 2.4.10 (Hamidoune[5]) *If $X = (V, E)$ is a nontrivial strongly connected vertex-transitive digraph, then $\kappa(X) \geq \lceil (\delta(X) + 1)/2 \rceil$.*

Proof.   Let A be an atom of X. Since $\delta(A) \leq |A| - 1$, we may use Corollary 2.4.7 and Proposition 2.4.9 to conclude that $\delta(X) \leq \delta(A) + \kappa(X) \leq |A| - 1 + \kappa(X) \leq 2\kappa(X) - 1$ and the result follows immediately. $\square$

Note that in the above proof, if X is antisymmetric then we have $\delta(A) \leq (|A| - 1)/2$. Thus we may improve the lower bound on $\kappa(X)$ in this case as follows.

Corollary 2.4.11 (Hamidoune[5]) *Let $X = (V, E)$ be a strongly connected vertex-transitive digraph. If X is antisymmetric, then $\kappa(X) \geq (2\delta(X) + 1)/3$.*

In the case of undirected graphs, which is to say symmetric digraphs, we may obtain the same lower bound as in Corollary 2.4.11. The reason for this is that in a symmetric digraph, the positive and negative neighborhoods of a set are the same. Thus the symmetric digraphs belong to the class of digraphs in which every atom is both a positive and negative fragment; we will say that a digraph with this property *has neutral atoms*. As we shall see later, every Abelian Cayley digraph has neutral atoms and thus all results proved for such digraphs apply to this class.

The next proposition is a generalization of a result proved by Mader [13] for undirected graphs and by Hamidoune [8] for Abelian Cayley digraphs.

Proposition 2.4.12 *Let* $X = (V, E)$ *be a strongly connected digraph having neutral atoms. If* W *is a minimum vertex disconnecting set and* A *an atom of* X, *then* A *is either disjoint from* W *or a subset of* W. *If* X *is vertex-transitive, then* W *is a disjoint union of atoms and thus* X *has at least 3 distinct atoms.*

Proof. Suppose $A \cap W \neq \emptyset$ and let $U \subset V$ be such that $N_X^+(U) = W$. Then A cannot be completely contained in either of the fragments U, $V \setminus (U \cup N_X^+(U)) = V \setminus (U \cup W)$. Since A is both a negative and positive atom, it must therefore be disjoint from both. It then follows that $A \subset W$. If X is vertex-transitive, then every vertex of W lies in an atom and thus W must be a disjoint union of atoms. The complement of W cannot consist of a single atom, since atoms induce strongly connected subdigraphs. Thus X must contain at least 3 atoms. $\square$

Proposition 2.4.13 *If* G *is a finite group and* $S \subset G \setminus \{1_G\}$ *is a normal set of generators for* G, *then* $X = Cay(G, S)$ *has neutral atoms. Moreover, the atom containing* $\{1_G\}$ *is a normal subgroup of* G *of cardinality at most* $|G|/3$.

Proof. If the atoms of X are vertices, which is to say $\kappa(X) = \delta(X)$, then the result is obvious. Thus we assume that the atoms of X are nontrivial. Let A be the atom containing the identity of G. We shall assume that A is a positive atom and prove that it is also a negative atom and that it is a normal subgroup of G. The proof in the case where A is a negative atom will then follow by considering the reverse digraph of X. Moreover, since X is vertex-transitive, it will follow that all atoms of X are both positive and negative atoms. From Corollary 2.4.5, A is the subgroup of G generated by $S \cap A$. Since A is a subgroup, $A^{-1} = A$. We then have $|A \cdot S^{-1}| = |(A \cdot S^{-1})^{-1}| = |S \cdot A^{-1}| = |S \cdot A| = |A \cdot S|$, where the last equality follows from the normality of S. Since A is nontrivial, $S \cap A \neq \emptyset$. If we choose an element s from $S \cap A$, then the elements s and s^{-1} both have right coset A, so that $A \subset AS \cap AS^{-1}$. It then follows that $\kappa(X) = |N^+(A)| =$

$|(A \cdot S) \setminus A| = |A \cdot S| - |A| = |(A \cdot S^{-1})| - |A| = |(A \cdot S^{-1}) \setminus A| = |N^-(A)|$, and we conclude that A is a negative atom. To see that A is a normal subgroup we note that by Proposition 2.2.2, right translations are automorphisms of X. Since inner automorphisms of G are compositions of left and right translations, they too are digraph automorphisms. Since an inner automorphism maps the identity to itself, it must also map the atom A onto itself. Therefore, A is preserved by all inner automorphisms of G and hence is a normal subgroup of G. Since X has neutral atoms it contains at least 3 distinct atoms and thus $|A| \leq |G|/3$.   □

We shall refer to a Cayley digraph of the form $\mathrm{Cay}(G, S)$ with S a normal set of generators of G as a *normally generated Cayley digraph*. We note that there are groups having no nontrivial normal subgroup of cardinality less than or equal to one-third the cardinality of the group. Thus any normally generated Cayley digraph over such a group is maximally connected. An important example of such a group is the symmetric group Sym_n on n letters. Thus we have as a corollary the following theorem of Imrich [11].

Theorem 2.4.14 *If X is a normally generated Cayley digraph over a symmetric group, then $\kappa(X) = \delta(X)$.*

We now prove some results for digraphs with neutral atoms which, of course, are then valid for normally generated Cayley digraphs and hence for the important class of Abelian Cayley digraphs. We also derive the fundamental Mader-Watkins lower bound for the connectivity of vertex-transitive undirected graphs. The results involving digraphs with neutral atoms were originally proved for undirected graphs and Abelian Cayley digraphs in the cited references.

Proposition 2.4.15 *Let $X = (V, E)$ be a nontrivial strongly connected, vertex-transitive digraph having neutral atoms. If A is an atom of X and $|A| = \kappa(X)$, then $N_X^+(A) \cap N_X^-(A) = \emptyset$.*

Proof. By Proposition 2.4.12, each of $N_X^+(A)$ and $N_X^-(A)$ is the disjoint union of atoms. Since atoms have cardinality $\kappa(X) = |N_X^+(A)| = |N_X^-(A)|$, both the positive and the negative neighborhoods of A are atoms. Thus they are either disjoint or equal. But if they were equal we would have $N^+(C^+(A)) = \emptyset$ and hence $V = A \cup N^+(A)$, which is impossible.   □

Since positive and negative neighborhoods coincide in undirected graphs, we thus have the following.

Corollary 2.4.16 *Let $X = (V, E)$ be a connected, vertex-transitive undirected graph. If A is an atom of X, then $\kappa(X) = m|A|$ for some integer $m \geq 2$ and thus $|A| \leq \kappa(X)/2$.*

We may now combine Corollaries 2.4.16 and 2.4.7 to obtain the following theorem, originally obtained independently by Mader and Watkins.

Theorem 2.4.17 (Mader[13], Watkins[16]) *If X is a connected vertex -transitive graph, then $\kappa(X) \geq (2\delta(X) + 1)/3$.*

We note that Theorem 2.4.17 does not hold if we weaken the condition of being undirected to the condition of having neutral atoms. To see this, one need only consider the Abelian Cayley digraphs $X = \mathrm{Cay}(\mathbf{Z}_n \times \mathbf{Z}_m, \mathbf{Z}_n \times \{0, 1\} \setminus \{(0, 0)\})$ with $m \geq 3, n \geq 2$. Each of these digraphs has $\delta = 2n - 1$ and $\kappa = n$. Note that these digraphs realize Hamidoune's lower bound for the connectivity of strongly connected vertex-transitive digraphs and do so with the least possible number of vertices.

Recall that a set U of vertices of a digraph $X = (V, E)$ is said to be *independent*, or *stable*, if no arc of X has both endpoints in U. The cardinality of a largest possible independent subset of a given subset T of V is denoted $\alpha(T)$. A digraph is said to be *regular* if every vertex of X has indegree and outdegree $\delta(X)$. The following is a miniscule extension of a lemma of Mader [14].

Proposition 2.4.18 *Let $X = (V, E)$ be a digraph. If the vertex set V can be partitioned into subsets each of which induces a regular subdigraph of X, then $\alpha(V) \leq |V|/2$.*

Proof. It is obvious that the proposition will follow if we can prove it for the case where X itself is regular. To that end, let S be an independent set of maximum cardinality. Then each arc initiating in S must terminate in $V \setminus S$ and thus we have $\delta|S| = \sum_{v \in S} d^+(v) \leq \sum_{v \in V \setminus S} d^-(v) = \delta|V \setminus S|$. $\square$

Since atoms of a vertex-transitive digraphs are themselves vertex-transitive, and hence regular, we may combine propositions 2.4.12 and 2.4.18 to obtain the following.

Corollary 2.4.19 *Let $X = (V, E)$ be a nontrivial strongly connected, vertex-transitive digraph having neutral atoms. If T is a minimum vertex disconnecting set of X, then $\alpha(T) \leq |T|/2$.*

Let us say that a digraph X is K_4-*free* if the symmetric digraph of X contains no clique on 4 vertices. We conclude this section with an investigation of the structure of strongly connected, vertex-transitive digraphs which have neutral atoms and are K_4-free. In particular, we show if such a digraph X has $\kappa(X) < \delta(X)$, then each atom has cardinality $\kappa(X)$. We then apply this result to show that no antisymmetric digraph falls in this class. The key to these results is the following lemma.

Lemma 2.4.20 (Hamidoune [9]) *Let $X = (V, E)$ be a K_4-free regular digraph and let $A \subset V$ be such that $\delta(A) > 0$. If $2|A| \leq |N^+(A)|$ and $\alpha(N^+(A)) \leq |N^+(A)|/2$, then $|N^+(A)| \geq \delta(X)$.*

Proof. For notational simplicity, let $\delta = \delta(X)$ and $k = |N^+(A)|$. Note that, by hypothesis, $|C^+(A)| = |A| + |N^+(A)| \leq 3k/2$. To show that $k \geq \delta$, we suppose to the contrary that $\delta + 1 \geq k$ and derive a contradiction. First we show that if (u, v) is an arc of X with $u, v \in A$, then $|N^+(u) \cap N^+(v)| \geq 2 + k/2$. Since X is K_4-free, there can be no arcs with both endpoints in the set $N^+(u) \cap N^+(v)$ and thus this set is an independent subset of V. Now $N^+(u) \cup N^+(v) \subset C^+(A)$ and so $3k/2 \geq |C^+(A)| \geq |N^+(u) \cup N^+(v)| = |N^+(u)| + |N^+(v)| - |N^+(u) \cap N^+(v)| = 2\delta - |N^+(u) \cap N^+(v)| \geq 2k + 2 - |N^+(u) \cap N^+(v)|$. We may thus deduce that $|N^+(u) \cap N^+(v)| \geq 2k + 2 - 3k/2 = 2 + k/2$, and the desired inequality is established.

Since $\delta(A) > 0$, we may choose an arc (u_1, u_2) with $u_1, u_2 \in A$. Since $\alpha(N^+(A)) \leq |N^+(A)|/2 = k/2$, there must be a vertex $u_3 \in N^+(u_1) \cap N^+(u_2) \cap A$. For $1 \leq i < j \leq 3$, let $B_{i,j} = N^+(u_i) \cap N^+(u_j)$; by the immediately preceding paragraph, we have $|B_{i,j}| \geq 2 + k/2$. Since X is K_4-free, $N^+(u_1) \cap N^+(u_2) \cap N^+(u_3) = \emptyset$ and thus the sets $B_{1,2}, B_{1,3}, B_{2,3}$ are pairwise disjoint. Therefore the cardinality of their union B is at least $6 + 3k/2$, which is impossible since B is a subset of the set $C^+(A)$ whose cardinality is at most $3k/2$. $\qquad\square$

Theorem 2.4.21 (Hamidoune[9]) *Let $X = (V, E)$ be a nontrivial strongly connected, vertex-transitive digraph which is K_4-free and has neutral atoms. If $\kappa(X) < \delta(X)$, then each atom of X has cardinality $\kappa(X)$.*

Proof. Let A be an atom of X. By Corollary 2.4.19, we deduce that $\alpha(N^+(A)) \leq |N^+(A)|/2$. Since $\kappa(X) < \delta(X)$, we know that X is not a complete symmetric digraph and thus may apply Proposition 2.4.12 to conclude that $|N^+(A)| = m|A|$ for some integer $m \geq 1$. But if $m \geq 2$, we may apply Lemma 2.4.20 to conclude that $\delta(X) \leq |N^+(A)| = \kappa(X)$, which contradicts our hypotheses. Thus we conclude that $m = 1$ and the proof is complete. $\square$

We may now apply Corollary 2.4.16 and Theorem 2.4.21 to deduce the following theorem of Mader [14].

Theorem 2.4.22 *If X is a connected undirected graph which is vertex-transitive and K_4-free, then $\kappa(X) = \delta(X)$.*

Theorem 2.4.23 (Hamidoune [9]) *Let $X = (V, E)$ be a nontrivial strongly connected, vertex-transitive digraph which is K_4-free and has neutral atoms. If X is antisymmetric, then $\kappa(X) = \delta(X)$.*

Proof. Suppose to the contrary that $\kappa(X) < \delta(X)$ and let A be an atom of X. From Theorem 2.4.21 we conclude that $|A| = \kappa(X)$ and thus from Proposition 2.4.15 that $N_X^+(A) \cap N_X^-(A) = \emptyset$. If we consider the symmetric digraph X_s of X, then $N_{X_s}^+(A) = N_X^+(A) \cup N_X^-(A)$ and thus $|N_{X_s}^+(A)| = |N_X^+(A)| + |N_X^-(A)| = 2\kappa(X) = 2|A|$. Note that X_s is also vertex-transitive, strongly connected and K_4-free. Moreover, a subset of V is independent in X if and only if it is independent in X_s. We also have $\alpha(N_{X_s}^+(A)) = \alpha(N_X^+(A) \cup N_X^-(A)) \leq \alpha(N_X^+(A)) + \alpha(N_X^-(A)) \leq |N_X^+(A)|/2 + |N_X^-(A)|/2 = |N_{X_s}^+(A)|/2$. Thus we may conclude from Lemma 2.4.20 that $2\delta(X) = \delta(X_s) \leq |N_{X_s}^+(A)| = 2\kappa(X)$, which contradicts our assumption that $\kappa(X) < \delta(X)$. $\square$

We remind the reader that one of our main motivations for proving the above results about digraphs with neutral atoms is that the results then hold for a large class of Cayley digraphs which are, after all, the main topic of this chapter.

REFERENCES

[1] D. Bauer, F. Boesch, C. Suffel and R. Tindell, Connectivity extremal problems and the design of reliable probablistic networks, The Theory and Applications of Graphs (Chartrand et. al., Editors), Wiley, New York (1985) 45-54.

[2] F. Boesch and J. Wang, Super line connectivity properties of circulant graphs, SIAM J. Alg. Discr. Methods 7 (1986) 89-98.

[3] G. Chaty, On critically and minimally k vertex (arc) connected digraphs, Proc. Keszthely (1976) 193-203.

[4] C. D. Godsil, Connectivity of minimal Cayley graphs, Arch. Math. 37 (1981) 437-476.

[5] Y. O. Hamidoune, Sur les atomes d'un graphe orienté, C.R. Acad. Sc. Paris Ser. A 284 (1977) 1253-1256.

[6] Y. O. Hamidoune, Quelques problèmes de connexité dans les graphes orienté, J. Comb. Theory Ser. B 30 (1981) 1-10.

[7] Y. O. Hamidoune, On the Connectivity of Cayley Digraphs, Europ. J. Comb. 5 (1984) 309-312.

[8] Y. O. Hamidoune, Sur la separation dans les graphes de Cayley Abelien, Disrcete Math. 55 (1985) 323-326.

[9] Y. O. Hamidoune, Connectivité des graphes de Cayley Abeliens sans K_4, Discrete Math. 83 (1990) 21-26.

[10] Y. O. Hamidoune and R. Tindell, Vertex transitivity and super line connectedness, SIAM J. Discr. Math. 3 (1990) 524-530.

[11] W. Imrich, On the connectivity of Cayley Graphs, J. Comb. Theory Series B 26 (1979) 323-326.

[12] W. Mader, Minimale n-fach kantenzusammenhängenden Graphen, Math. Ann. 191 (1971) 21-28.

[13] W. Mader, Ein Eigenschaft der Atome endlicher Graphen, Arch. Math.22 (1971) 331-336.

[14] W. Mader, Über den zusammen symmetricher Graphen, Arch. Math. 21 (1970) 331-336.

[15] R. Tindell, Edge connectivity properties of symmetric graphs, preprint, Stevens Institute of Technology 1982.

[16] M. E. Watkins, Connectivity of transitive graphs, J. Comb. Theory 8 (1970) 23-29.

3

DE BRUIJN DIGRAPHS, KAUTZ DIGRAPHS, AND THEIR GENERALIZATIONS

Ding-Zhu Du
Feng Cao
University of Minnesota

D. Frank Hsu
Fordham University

3.1   INTRODUCTION

An interesting problem in network designs is as follows: Given natural numbers n and d, find a digraph (directed graph) with n vertices, each of which has outdegree at most d, to minimize the diameter and to maximize the connectivity. This is a multiobjective optimization problem. Usually, for such a problem, solution is selected based on tradeoff between two objective fuctions. However, for this problem, it is different; that is, there exists a solution which is optimal or nearly optimal to both. Such a solution comes from study of de Bruijn digraphs, Kautz digraphs, and their generalizations. In this chapter, we introduce and survey results on these subjects.

To begin with, we study de Bruijn digraphs and Kautz digraphs in this section. Then we study their generalizations and diameter, line-connectivity, super-line-connectivity, connectivity, and Hamiltonian property of the generalizations in the other six sections.

3.1.1   de Bruijn Digraphs

The de Bruijn digraph was first discovered for solving a coding problem many years ago [8]. Later, one found that it has many nice properties with many

Ding-Zhu Du and D. Frank Hsu (eds.), Combinatorial Network Theory, 65–105.
© 1996 *Kluwer Academic Publishers. Printed in the Netherlands.*

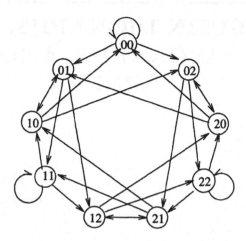

Figure 3.1 de Bruijn digraph $B(3,2)$.

applications. The de Bruijn digraph can be defined in several ways. The following is a popular one.

Definition 3.1.1 *Let X be the set of vectors x with t components in $\{0, 1, \ldots, d-1\}$. de Bruijn digraph $B(d,t)$ is the digraph with vertex set X such that an edge exists from $(x_1, \cdots, x_t)$ to $(y_1, \cdots, y_t)$ iff $x_2 = y_1$, $x_3 = y_2$, $\cdots$, $x_t = y_{t-1}$.*

For example, de Bruijn digraph $B(3,2)$ is shown in Figure 3.1.

From the definition, it is easy to see that de Bruijn digraph $B(d,t)$ has d^t vertices and diameter t. Each vertex has outdegree d and indegree d.

A digraph is called *d-regular* if every vertex in the digraph has the same out-degree d. de Bruijn digraph $B(d,t)$ is d-regular.

Plesnik and Znam [40] showed that a digraph with outdegree at most d and diameter D can have at most $d + d^2 + \cdots + d^D$ vertices. It follows that a digraph with n vertices and outdegree at most d has diameter D satisfying inequality

$$n < (d^{D+1} - 1)/(d - 1),$$

that is,

$$D \geq \lfloor \log_d(n(d-1) + 1) \rfloor. \tag{3.1}$$

Thus, a digraph with d^t vertices each of outdegree at most d has diameter at least t. This means that de Bruijn digraph $B(d, t)$ achieves the minimum value of diameter for digraphs with d^t vertices each of outdegree at most d.

Note that de Bruijn digraph $B(d, t)$ is d-regular and moreover, some vertices (such as $(0, 0, \cdots, 0)$) contain loops. Therefore, its connectivity is at most $d-1$. M. Imase, T. Soneoka and K. Okada [33] proved that de Bruijn digraph $B(d, t)$ is exactly $(d-1)$-connected and moreover, there exist $(d-1)$ vertex-disjoint paths between any pair of vertices in $B(d, t)$, one of length at most t and $d-2$ of length at most $t+1$. To show their result, let us first prove two lemmas.

The first lemma states an important property, which was given by M.A. Fiol, J.L.A. Yebra and I. Alegre [23] for an alternative definition of the de Bruijn digraph.

Lemma 3.1.2 $B(d, 1)$ *is the complete digraph on* d *vertices with a loop at each vertex. For* $t \geq 2$, $B(d, t)$ *is the line digraph of* $B(d, t-1)$.

Proof. Let ψ be the map from the edge set of $B(d, t-1)$ to the vertex set of $B(d, t-1)$ defined by mapping $(x_1, \ldots, x_{t-1}) \to (x_2, \ldots, x_{t-1}, i)$ to $(x_1, \ldots, x_{t-1}, i)$. It is easy to verify that ψ is an isomorphism from the line digraph of $B(d, t-1)$ to $B(d, t)$. □

Lemma 3.1.3 *Let* u *and* v *be two distinct vertices in de Bruijn digraph* $B(d, t)$ *such that there is an edge from* u *to* v. *Then there exist* $d-2$ *vertex-disjoint paths of length at most* $t+1$ *from* u *to* v, *not passing through edge* (x, y).

Proof. Let $u = (x_1, \ldots, x_t)$ and $v = (x_2, \ldots, x_t, \alpha)$. Consider the following paths P_β:

$$(x_1, \ldots, x_t), (x_2, \ldots, x_t, \beta), \ldots, (\beta, x_2, \ldots, x_t), (x_2, \ldots, x_t, \alpha) \qquad (3.2)$$

where $\beta \neq \alpha, x_1$. Note that these are $d-2$ paths; each may not be simple, but, two endpoints appear only once. For each β, every vertex in path P_β other than two endpoints has coordinates forming the same multi-set $\{x_2, \cdots, x_t, \beta\}$. For different β, the corresponding multi-sets are different. Therefore, these $d-2$ paths are vertex-disjoint. Clearly, their lengths are exactly $t+1$. □

Theorem 3.1.4 *There exist* $(d-1)$ *vertex-disjoint paths between any pair of vertices in* $B(d, t)$, *one of length at most* t *and* $d-2$ *of length at most* $t+1$.

Proof. We prove it by induction on t. It is trivial for $t = 1$ since $B(d, 1)$ is a complete digraph on d vertices each having a loop. For $t \geq 2$, we notice that $B(d, t)$ is the line-digraph of $B(d, t - 1)$. Consider any pair of vertices u and v in $B(d, t)$. Suppose in $B(d, t - 1)$ that edge u goes to vertex a and that edge v comes from vertex b. If $a \neq b$, then by the induction hypothesis, $B(d, t - 1)$ has $(d - 1)$ vertex-disjoint paths from a to b, one of length at most $t - 1$ and $d - 2$ of length at most t. These $d - 1$ paths induce $d - 1$ vertex-disjoint paths from u to v in $B(d, t)$, one of length at most t and $d - 2$ of length at most $t + 1$. If $a = b$, then $B(d, t)$ has an edge from u to v. By the above lemma, $B(d, t)$ has $d - 2$ edge-disjoint paths of length at most $t + 1$ from u to v, not passing edge (u, v). Moreover, edge (u, v) gives the $(d - 1)$st path of length at most t from u to v. □

From the above results, we know that de Bruijn digraph $B(d, t)$ achieves the minimum diameter and nearly the maximum connectivity (one different from the maximal value). The connectivity is hurt by the existence of loops. A natual improvement is to replace all loops by a cycle connecting all loop-vertices. Does this really give a better connectivity for de Bruijn digraph? The answer is "Yes". We will prove this in section 3.

Note that the line digraph of a digraph is Hamiltonian iff the digraph is Eulierian. Since the indegree equals the outdegree at each vertex in any $B(d, t)$, every connected $B(d, t)$ is Eulierian. Therefore, as a corollary of its connectivity, we have that every $B(d, t)$ with $d \geq 2$ is Hamiltonian. Hence, every de Bruijn digraph $B(d, t)$ for $d \geq 2$ has a Hamiltonian circuit.

3.1.2   Kautz Digraphs

Every Kautz digraph is a subgraph of a de Bruijn digraph. It is defined as follows.

Definition 3.1.5 *Let Y be the set of vectors with t components in $\{0, 1, \ldots, d\}$ satisfying that any two adjacent components are different. Kautz digraph $K(d, t)$ is the digraph with vertex set Y such that an edge exists from $(x_1, \cdots, x_t)$ to $(y_1, \cdots, y_t)$ iff $x_2 = y_1, x_3 = y_2, \cdots, x_t = y_{t-1}$.*

From the definition, it is clear that Kautz digraph $K(d, t)$ can be obtained from de Bruijn digraph $B(d + 1, t)$ by removing all vertices with equal adjacent components.

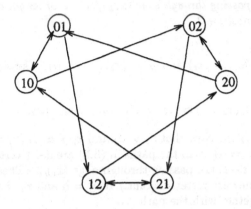

Figure 3.2   Kautz Digraph $K(2,2)$

If a vertex $(x_1, \cdots, x_t)$ in de Bruijn digraph $B(d+1, t)$ has a loop, then we must have $x_1 = x_2 = \cdots = x_t$. Thus, Kautz digraphs have no loop. This property enable them to have better connectivity than that of de Bruijn digraphs. To show this, we first prove two lemmas.

Lemma 3.1.6 *Kautz digraph $K(d, 1)$ is a complete symmetric digraph of order $d + 1$. For $t > 1$, Kautz digraph $K(d, t)$ is the line graph of Kautz digrap $K(d, t - 1)$.*

Proof. Similar to the proof of Lemma 3.1.2.   □

Note that both indegree and outdegree of each vertex in Kautz digraph $K(d, t)$ are d. It follows that the connectivity of Kautz digraph $K(d, t)$ is at most d. In fact, Kautz digraph $K(d, t)$ has been proved to be d-connected. In the other words, it achieves the maximum connectivity. More precisely, M. Imase, T. Soneoka and K. Okada [33] proved that there exist d vertex-disjoint paths between any pair of vertices in $K(d, t)$, one of length at most t, $d - 3$ of length at most $t + 1$ and two of length at most $t + 2$, and S.M. Reddy, J.G. Kuhl, S.H. Hosseini and H. Lee [41] proved that there exist d vertex-disjoint paths between any pair of vertices in $K(d, t)$, $d - 1$ of length at most $t + 1$ and one of length at most $t + 2$. Next, we show a result a little stronger than the above two.

Lemma 3.1.7 *Let x and y be two distinct vertices in the Kautz digraph $K(d, t)$ such that there is an edge from x to y. There exist $d - 1$ vertex-disjoint paths* *from x to y, not passing through edge (x,y), $d-2$ of length at most $t+1$ and one of length at most $t+2$.*

Proof. Let $x = (x_1,\ldots,x_t)$ and $y = (x_2,\ldots,x_t,\alpha)$. Consider the following paths:

$$(x_1,\ldots,x_t),(x_2,\ldots,x_t,\beta),\ldots,(\beta,x_2,\ldots,x_t),(x_2,\ldots,x_t,\alpha) \qquad (3.3)$$

where $\beta \neq \alpha, x_1, x_2, x_t$. Note that $x_1 \neq x_2$ and $x_t \neq \alpha$. If $(x_1 = \alpha$ and $x_2 = x_t)$ or $(x_1 = x_t$ and $x_2 = \alpha)$, then the paths in (3.3) are $d-1$ vertex-disjoint paths of length at most $t+1$, not passing through edge (x,y). (These paths may not be simple. But, they are vertex-disjoint.) If $x_1 = \alpha$ and $x_2 \neq x_t$, then the $d-2$ paths in (3.3) together with the path

$$(x_1,\ldots,x_t),(x_2,\ldots,x_t,x_2),\ldots,(x_t,x_2,\ldots,x_t),(x_2,\ldots,x_t,\alpha), \qquad (3.4)$$

form $d-1$ vertex-disjoint paths of length at most $t+1$, not passing through edge (x,y). If $x_1 \neq \alpha$ and $x_2 = x_t$, then the $d-2$ paths in (3.3) together with the path

$$(x_1,\ldots,x_t),(x_2,\ldots,x_t,x_1),(x_3,\ldots,x_t,x_1,\alpha),(x_4,\ldots,x_t,x_1,\alpha,x_2),$$
$$\ldots,(x_2,\ldots,x_t,\alpha), \qquad (3.5)$$

form $d-1$ vertex-disjoint paths, $d-2$ of length at most $t+1$ and one of length at most $t+2$, not passing through edge (x,y). If $x_1 \neq x_t$ and $x_2 = \alpha$, then the $d-2$ paths in (3.3) together with the path

$$(x_1,\ldots,x_t),(x_2,\ldots,x_t,x_1),(x_3,\ldots,x_t,x_1,x_t),(x_4,\ldots,x_t,x_1,x_t,x_2),$$
$$\ldots,(x_2,\ldots,x_t,\alpha),$$

form $d-1$ vertex-disjoint paths, $d-2$ of length at most $t+1$ and one of length at most $t+2$, not passing through edge (x,y). If $x_1 = x_t$ and $x_2 \neq \alpha$, then the $d-2$ paths in (3.3) together with the path

$$(x_1,\ldots,x_t),(x_2,\ldots,x_t,x_2),(x_3,\ldots,x_t,x_2,\alpha),(x_4,\ldots,x_t,x_2,\alpha,x_2),$$
$$\ldots,(x_2,\ldots,x_t,\alpha),$$

form $d-1$ vertex-disjoint paths, $d-2$ of length at most $t+1$ and one of length at most $t+2$, not passing through edge (x,y). If $x_1 \neq \alpha$, $x_1 \neq x_t$, $x_2 \neq x_t$, and $x_2 \neq \alpha$, then the $d-3$ paths in (3.3) together with the paths in (3.4) and the path in (3.5) form $d-1$ vertex-disjoint paths, $d-2$ of length at most $t+1$ and one of length at most $t+2$, not passing through edge (x,y). $\qquad \square$

Theorem 3.1.8 *Let x and y be any two vertices in $K(d,t)$. Then there exist d vertex-disjoint paths from x to y, one of length at most t, $d-2$ of length at most $t+1$, and one of length at most $t+2$.*

Proof. We use induction on t. The theorem is true for $t = 1$ as $K(d,1)$ is the complete symmetric digraph on $d + 1$ vertices. For $t \geq 2$, assume that the theorem is true for $K(d,i)$, $i \leq t - 1$.

Any two vertices x and y in $K(d,t)$ correspond to two edges in $K(d,t-1)$. Let these two edges be $x = (x_0, x_1)$ and $y = (y_0, y_1)$. If $x_1 \neq y_0$, then by the induction hypothesis, we have d vertex-disjoint paths from vertex x_1 to y_0 in $K(d,t-1)$, one of length at most $t - 1$, $d-2$ of length at most t, and one of length at most $t + 1$. These d paths induce d vertex-disjoint paths in $K(d,t)$ from x to y, one of length at most t, $d-2$ of length at most $t + 1$, and one of length at most $t + 2$. If $x_1 = y_0$, then there exists an edge from x to y and by the above lemma, there are $d - 1$ vertex-disjoint paths from x to y, not passing through edge (x,y), $d - 2$ of length at most $t + 1$ and one of length at most $t + 2$. Thus, the theorem is true for $K(d,t)$. $\square$

It is easy to see from the above theorem that $K(d,t)$ has diameter at most t. Moreover, from (3.1), we know that a d-regular digraph with $d^t + d^{t-1}$ vertices has diameter at least

$$\lfloor \log_d \left((d^t + d^{t-1})(d-1) + 1 \right) \rfloor = t$$

(Note: $d^t \leq d^{t+1} - d^{t-1} + 1 = (d^t + d^{t-1})(d-1) + 1 < d^{t+1}$.) Therefore, $K(d,t)$ also achieves the minimum diameter.

3.2 GENERALIZATIONS

We already see that de Bruijn digraphs and Kautz digraphs are nearly-optimal or optimal for our problem. Unfortunately, they exist only for some special numbers of vertices. To provide optimal or nearly-optimal solutions in general case, we introduce some generalizations of de Bruijn and Kautz digraphs in this section.

3.2.1   Generalized de Bruijn Digraphs

de Bruijn digraphs were generalized by M. Imase and M. Itoh [30] and S.M. Reddy, D.K. Pradhan and J.G. Kuhl [42], independently. To see this generalization, let us first look at de Bruijn digraph $B(d,t)$ in a different way. Suppose $n = d^t$. Define a one-one and onto map ψ from the vertex set of $B(d,t)$ to Z_n by $\psi(x_1, \ldots, x_t) = x_1 d^{t-1} + x_2 d^{t-2} + \cdots + x_t$. The map ψ give a new name for each vertex of $B(d,t)$. For this new name, the edge from i to j exists iff $j \equiv di + k \pmod{n}$ for some $k = 0, \ldots, d - 1$.

Definition 3.2.1 *Let n and d be two natural numbers satisfying $d < n$. Generalized de Bruijn digraph $G_B(d,n)$ is a digraph with n vertices labeled by the residues modulo n such that an edge from i to j exists iff $j \equiv di + k \pmod{n}$ for some $k = 0, \ldots, d - 1$.*

A subset of vertices is called a *consecutive run* if the vertices in the subset are numbered consecutively by integers modulo n. Clearly, in $G_B(d,n)$, all edges from a consecutive run of size m would reach a consecutive run of size $\min(n, dm)$. It follows that $G_B(d,n)$ has diameter at most $\lceil \log_d n \rceil$ which differs at most one from the lower bound $\lfloor \log_d(n(d-1) + 1) \rfloor$.

M. Imase, T. Soneoka and K. Okada [32] proved that the generalized de Bruijn digraph is $(d-1)$-connected. Du, Hsu and G. W. Peck [19] proved that $G_B(d,n)$ can be modified to be d-connected. This result was independently obtained by T. Soneoka, H. Nakada, M. Imase and Y. Manabe [44] [45]. It follows from these results that modified generalized de Bruijn digraphs have nearly minimum diameter and maximum connectivity.

3.2.2   Imase-Itoh Digraphs

Kautz digraphs were first generalized by I. Imase and M. Itoh [31].

Definition 3.2.2 *Let n and d be two natural numbers satisfying $d < n$. Imase-Itoh digraph $G_I(d,n)$ is a digraph with n vertices labeled by the residues modulo n such that an edge from i to j exists iff $j \equiv -d(i + 1) + k \pmod{n}$ for some $k = 0, \ldots, d - 1$.*

To see that $G_I(d,n)$ is a generalization of $K(d,t)$, let us first prove a property of $G_I(d,n)$.

Lemma 3.2.3 $G_I(d,dn)$ *is isomorphic to the line digraph of* $G_I(d,n)$.

Proof. Map edge $i \to -d(i+1)+k$ of $G_I(d,n)$ to vertex $-d(i+1)+k$ of $G_I(d,dn)$. $\qquad\qquad\qquad\qquad\qquad\qquad\qquad\qquad\qquad\qquad\qquad\qquad\quad\square$

Note that $G_I(d,d+1)$ is a complete digraph of $d+1$ vertices which is isomorphic to $K(d,1)$. It is easy to see that $G_I(d,d^{t-1}(d+1))$ is isomorphic to $K(d,t)$.

It is similar to the generalized de Bruijn digraph, Imase-Ito digraph $G_I(d,n)$ has diameter at most $\lceil \log_d n \rceil$. Moreover, Imase and Itoh [31] proved that if $n = d^s + d^{s-t}$ for some odd t and positive integer $s \geq t$, then $G_I(d,n)$ has diameter $\lceil \log_d n \rceil - 1$ which is the minimum.

While the diameter of $G_I(d,n)$ was determined easily, the connectivity was left open for a while. M. Imase, T. Soneoka and K. Okada [32] proved that $G_I(d,n)$ is at least $(d-1)$-connected. D.-Z. Du and F.K. Hwang [15] proved that the line-connectivity of $G_I(d,n)$ is d if and only if $d+1$ divides n. This also implies that $G_I(d,n)$ is d-connected when $d+1$ and d both divide n. N. Homobono and C. Peyrat [27] proved that $G_I(d,n)$ is d-connected when $d+1$ divides n, d and n are not relatively prime, and $n \geq d^5$.

The above results indicate that $G_I(d,n)$ is also a good solution for our problem.

3.2.3 Consecutive-d Digraphs

To give a uniform treatment for generalized de Bruijn digraphs and Imase-Itoh digraph, Du, Hsu and Hwang [12] proposed the concept of consecutive-d digraph.

Definition 3.2.4 *Let n and d be two natural numbers satisfying $d < n$. Given $q \in Z_n \setminus \{0\}$ and $r \in Z_n$, consecutive-d digraph $G(d,n,q,r)$ is defined to be a digraph with n $(> d)$ vertices labeled by the residues modulo n such that an edge from i to j exists iff $j \equiv qi + r + k \pmod{n}$ for some $k = 0, \ldots, d-1$.*

When $q = d$ and $r = 0$, $G(d, n, q, r)$ is generalized de Bruijn digraph $G_B(d, n)$. When $q = -d = r$, $G(d, n, q, r)$ is Imase-Ito digraph $G_I(d, n)$.

In this subsection, we show some basic properties of $G(d, n, q, r)$.

Let $\lambda = gcd(q - 1, n)$. (note: $\lambda = n$ if $q = 1$.) Denote by $(x)_n$ the residue of x modulo n, represented by a number in $\{0, 1, \ldots, n - 1\}$. An edge is said to be with k-value i if it is contained in the subgraph $G(1, n, q, r + i)$. The following lemma is about the distribution of loops in the consecutive-d digraph.

Lemma 3.2.5 $G(d, n, q, r)$ *has the following properties:*

(a) *Each node has at most one loop.*

(b) *If $d \geq 2$ then $G(d, n, q, r)$ has either no loop or at least two loops.*

(c) *$G(d, n, q, r)$ has no loop iff $d < \lambda$ and $(r)_\lambda \leq \lambda - d$.*

(d) *If $d \leq \lambda$, then all loops of $G(d, n, q, r)$ are with the same k-value.*

(e) *If $\lambda = 1$, then there exists exactly one loop with each k-value. If $\lambda > 1$, then either there is no loop or there are exactly λ loops with each k-value.*

(f) *If $|q-1| \leq d$ and x is a loop-node, then either $x + \lfloor n/(q-1) \rfloor$ or $x + \lceil n/(q-1) \rceil$ is a loop-node.*

Proof. To prove (a), (b), and (c), note that there is a loop at node i iff for some $k \in \{0, \ldots, d - 1\}$,

$$(q - 1)i + r + k \equiv 0 \pmod{n}. \tag{3.6}$$

For each i, there exists at most one k-value such that (3,6) holds. Thus, (a) is true. To see (c), note that $G(d, n, q, r)$ has no loop iff $r + k$ is not divisible by λ for all k. This happens iff $d < \lambda$ and $(r)_\lambda \leq \lambda - d$.

For (d), note that when $d \leq \lambda$, there exists at most one k-value such that $r + k$ is divisible by λ. Therefore, loops could exist only for such a k-value.

To see (e), note that if $\lambda = 1$, the equation (3.6) has exactly one solution i for each k; if $\lambda > 1$, then $i, i + n/\lambda, \cdots, i + (\lambda - 1)n/\lambda$ satisfy (3.6) simultaneously.

(b) follows immediately from (e) and (f).

Finally, we prove (f). If $q = 1$, then $\lambda = n$. So, either every node has a loop or every node has no loop. Thus, (f) holds trivially in this case. Next, consider $q \neq 1$. Suppose that the loop at x is $x \to qx + r + k$. Denote $k(x) = qx + r + k$. Then $k(x + 1) - (x + 1) = k(x) - x + (q - 1)$. Denote $y = \lfloor n/(q - 1) \rfloor$ and $z = \lceil n/(q - 1) \rceil$. If $y = z$, then $k(x + y) - (x + y) = k(x) - x + n$, so that $x + y$ is a loop-node. If $y \neq z$, then $z = y + 1$ and $(q - 1)y < n < (q - 1)z$ (if $q > 0$) or $(q - 1)y > n(q - 1)z$ (if $q < 0$). Since $|q - 1| \leq d$, we have either $k + n - (q - 1)y$ or $k + n - (q - 1)z$ is in between 0 and $d - 1$. It follows that either $x + y$ or $x + z$ has a loop. □

Denote $g = gcd(n, q)$. In $G(d, n, q, r)$, every vertex has outdegree d. However, indegrees may not be equal at different vertices. The next two lemmas provide the properties of indegrees.

Lemma 3.2.6 *The indegree of every vertex of $G(d, n, q, r)$ is divisible by g.*

Proof. Partition the n vertices of $G(d, n, q, r)$ into $n' = n/g$ groups of g vertices where the group $\bar{i}$ consists of vertices $\{i, i + n', \ldots, i + (g - 1)n'\}$. Then the vertices in the same group have the same set of d successors. Therefore the indegree of each vertex is a multiple of g. □

Lemma 3.2.7 *The indegree of each vertex of $G(d, n, q, r)$ is d if and only if $g \mid d$.*

Proof. Consider any vertex j. The edge $i \to j$ exists iff for some $k = 0, \ldots, d-1$, $j \equiv qi + r + k \pmod{n}$. This equation has a solution iff g divides $r + k - j$. Since g divides d, there are d/g values for k such that g divides $r + k - j$. For each of such k's, the equation is equivalent to the following.

$$0 \equiv q'i + k' \pmod{n'}$$

where $n' = n/g, q' = q/g$ and $k' = (r + k - j)/g$. Since $gcd(n', q') = 1$, there are exactly d/g numbers in $\{0, \ldots, d - 1\}$ satisfying (3,7). Therefore, the indegree of vertex j is d. □

The generalization of Lemmas 3.1.2 and 3.1.6 is given as follows.

Lemma 3.2.8 *If d divides n then $G(d, n, q, r)$ is the line-graph of $G(d, n/d, q, r)$.*

Proof. Consider a digraph $\bar{G}$ with vertices $g_0, \ldots, g_{n'-1}$ where $n' = n/d$, and with edges labeled by $0, \ldots, n$; each vertex g_i has in-edges $i, i+n', \ldots, i+(d-1)n'$ and out-edges $qi+r, qi+r+1, \ldots, qi+r+d-1$. Clearly, there is an edge from g_i to g_j iff $j \equiv qi + r + k \pmod{n'}$ for some $k = 0, \ldots, d-1$. Thus, $\bar{G}$ is isomorphic to $G(d, n', q, r)$. On the other hand, the line graph of $\bar{G}$ is $G(d, n, q, r)$. □

It is possible that two consecutive-d digraphs with different parameters are isomorphic each other. The following is an example.

Lemma 3.2.9 *If $r \equiv r' (mod\ h)$ where $h = gcd(n, q-1)$, then $G(d, n, q, r)$ and $G(d, n, q, r')$ are isomorphic each other.*

Proof. Write $r = hx + r'$. Let y be a solution of the equation $(q-1)y \equiv hx \pmod{n}$. It is easy to verify that the map $f : i \to i + y$ will give an isomorphism from $G(d, n, q, r)$ onto $G(d, n, q, r')$. □

By Lemma 3.2.9, we can assume throughout that $0 \leq r \leq gcd(n, q-1)$.

3.3   DIAMETER

We study the diameter of $G(d, n, q, r)$ in this section.

Let q' be the magnitude of q when we write q as an integer between $-n/2$ and $n/2$. Recall that a subset of vertices is called a consecutive run if the vertices in the subset are numbered consecutively by integers modulo n.

Theorem 3.3.1 *If $d \geq q' > 1$ then the diameter of $G(d, n, q, r)$ is at most $\lceil \log_{q'} n \rceil$.*

Proof. We note that all edges coming from consecutive run of m vertices can reach a consecutive run of size at least $\min(n, q'm)$ vertices. Therefore, all paths of length t starting from a vertex can reach at least $\min((q')^t, n)$ consecutively numbered vertices. Therefore, its diameter is bounded by $\lceil \log_{q'} n \rceil$. □

Imase and Itoh [31] showed that the digraph $G_I(d, n)$ for $n = d^s + d^{s-t}$ with t odd and $t \leq s$ has the minimum diameter for digraphs with n vertices each of outdegree at most d. The following is a variation of their result.

Theorem 3.3.2 *If $n = d^s + d^{s-t}$ with t odd and $t \leq s$, then the diameter of $G(d, n, -d, r)$ is $\lceil \log_d n \rceil - 1$.*

Proof. Note that all paths of length s starting from a vertex i will reach a consecutive run with following two endpoints

$$(-d)^s i + \frac{(-d)^s - 1}{d + 1} \cdot r$$

and

$$(-d)^s i + \frac{(-d)^s - 1}{d + 1}(r + d - 1).$$

Since the given conditions imply that $(-d)^s \equiv (-d)^{s-t} \pmod{n}$, the above two endpoins are also the endpoints of the consecutive run reached by the paths of length $s - t$ starting from the vertex i. These two consecutive runs contain, respectively, d^s and d^{s-t} vertices. This means that they just cover Z_n. Therefore, the diameter of the digraph is at most s. However, it is easy to see that the diameter is at least s. Finally, the proof is completed by noting that $s = \lceil \log_d n \rceil - 1$. $\qquad\qquad\qquad\qquad\qquad\qquad\qquad\qquad\qquad\qquad\qquad\qquad\square$

Although many consecutive-d digraphs do not achieve the minimum diameter for given d and n, we believe that there exists one achieving the minimum. Precisely speaking, we conjecture that for any d and n, there exists a generalized de Bruijn-Kautz digraph $G(d, n, q, r)$ whose diameter is less than or equal to that of any digraph with n vertices each of indegree and outdegree at most d.

The s-diameter vulnerability $D(s; G)$ of a digraph G is the maximum of diameters of subgraphs obtained by removing s arbitrary vertices from G. For example, $D(0; G)$ is the diameter of G. For de Bruijn digraphs and Kautz digraphs, the s-diameter vulnerability is as follows [16, 33]:

$$
\begin{aligned}
D(0; B(d, t)) &= t \\
D(s; B(d, t)) &= t + 1 \text{ for } 1 \leq s \leq d - 2 \\
D(s; B(d, t)) &= 0 \text{ for } s \geq d - 1,
\end{aligned}
$$

and

$$
\begin{aligned}
D(0; K(d, t)) &= t \\
D(s; K(d, t)) &= t + 1 \text{ for } 1 \leq s \leq d - 2 \\
D(d - 1; K(d, t)) &= t + 2 \\
D(s; K(d, t)) &= 0 \text{ for } s \geq d.
\end{aligned}
$$

However, it is still an open problem of determining the s-diameter vulnerability of consecutive-d digraphs.

3.4   LINE CONNECTIVITY

We study the line-connectivity of consecutive-d digraphs and their modifications in this section.

Du, Hsu, and G.W. Peck [19] showed the following.

Theorem 3.4.1 *Let $d \geq 2$. $G(d, n, q, r)$ is d-line-connected iff it has no loop and every node has indegree d. Moreover, if every node has indegree d then $G(d, n, q, r)$ is at least $(d - 1)$-line-connected..*

This result indicates that consecutive-d digraphs have very good line-connectivity, especially if no loop exists.

How do we modify consecutive-d digraphs to reach the maximum line-connectivity if loops exist? Essentially, the modification is to replace all loops by some cycles. A digraph is called a *modified* $G(d, n, q, r)$ if it is constructed from $G(d, n, q, r)$ by connecting all loop-nodes into disjoint cycles of size at least two and deleting all loops. The modification is said to be *cyclic* if all loops are replaced by a single cycle. The modified $G(d, n, q, r)$ is said to be *simple* if there is no multiple edge in the graph. Not every consecutive-d digraph has a simple and cyclic modification. For example, all cyclic modifications of $G(3, 4, 3, 0)$ must contains a multiple edges. Since multiple edges contribute nothing to diameter, avoiding multiplicity is of some interests.

The next two theorems were proved by Du, Hsu, and Kleitman [18].

Theorem 3.4.2 *Let $g = gcd(n, q)$ and $d \geq 2$. There exists a modified $G(d, n, q, r)$ having line-connectivity d if and only if g divides d. If g divides d, then the following holds:*

(a) Every cyclically modified $G(d, n, q, r)$ is d-line-connected.

(b) Every modified $G(d, n, q, r)$ is d-line-connected unless the following occurs

(b1) $g = 1$, $d \leq 3$ and $q \equiv \pm 1 \pmod{n}$, or

(b2) $G(d, n, q, r)$ is isomorphic to $G(3, 6, 3, 1)$, or

(b3) $g = d = 2$ and $n = ((q)_n + 1)2^s$ for some natural number s.

Theorem 3.4.3 *Let q^* be the magnitude of q. Suppose g divides d. Then for $d \geq 2$ and $n > (q^* + 1)d$, there always exists a d-line-connected simple modified $G(d, n, q, r)$.*

In the following, we prove the above results.

3.4.1   Edge-Cuts

The proofs of Theorems 3.4.1, 3.4.2, and 3.4.3 are based on investigation about edge-cuts of size $d - 1$. If $G(d, n, q, r)$ is not d-line-connected, then we must have an edge-cut of size $d - 1$. That is, the nodes can be partitioned into two disjoint nonempty sets A and B such that at most $d - 1$ edges go from A to B. When g divides d, the indegree and the outdegree are equal at each node. Thus, the number of edges from B to A equals the number of edges from A to B. This will always hold in the following.

Lemma 3.4.4 *Suppose that the nodes of $G(d, n, q, r)$ can be partitioned into two disjoint nonempty sets A and B such that at most $d - 1$ edges go from A to B. If $d \geq 2$ and $g = 1$, then A and B both are consecutive runs unless $G(d, n, q, r)$ is isomorphic to $G(3, 4, 1, 0)$ or $G(3, 4, -1, 0)$.*

Proof. Without loss of generality, assume $|A| \leq |B|$. If $|A| = 1$ then the conclusion holds trivially. Next, we consider $|A| \geq 2$. Suppose that either A or B does not consist of consecutive nodes. Then, we can find at least four different consecutive pairs of nodes, $\{a, a+1\}$, $\{b, b+1\}$, $\{c, c+1\}$ and $\{e, e+1\}$ such that in each pair, one node is in A and another is in B. Since $|A| \geq 2$, we have $n \geq 4$. Thus, at least two consecutive pairs are disjoint. We now consider a pair $\{a, a + 1\}$. Note that each node has outedges to d consecutive nodes. Since $g = 1$, there is exactly one node linking to any d consecutive nodes. Thus, there are exactly $d - 1$ nodes going to both a and $a + 1$. Among these $2(d - 1)$ edges, exactly $d - 1$ edges going between A and B. Since at most $2(d-1)$ edges go between A and B and $d \geq 2$, there are at most two disjoint pairs among the four. Without loss of generality, assume that the first two pairs are joint with $b \equiv a + 1 \pmod{n}$. Since there are exactly $d - 2$ nodes having edges going to all of a, $a + 1$ and $a + 2$, there are exactly d nodes linking to either both a and $a + 1$ or both $a + 1$ and $a + 2$. This contributes at least d edges going between A and B. Since there are at most $2(d - 1)$ edges between A and B, each of other two pairs cannot be disjoint from $\{a, a + 1, a + 2\}$. This happens only if $\{c,e\} = \{(a-1)_n, (a+2)_n\}$. In this case, we must also have $a + 3 \equiv a - 1$ (mod n) since, otherwise, $\{a-1, a\}$ and $\{a+1, a+2, a+3\}$ are disjoint. $a + 3 \equiv a - 1$ (mod n) implies $n = 4$. For $n = 4$, there are few cases so that we can find the exceptions easily. $\qquad\qquad\qquad\qquad\qquad\qquad\qquad\qquad\qquad\square$

Lemma 3.4.5 *Let $d \geq 2$ and $g = 1$. Suppose that the nodes of $G(d, n, q, r)$ can be partitioned into disjoint nonempty sets A and B such that at most $d - 1$ edges go from A to B. Then either A or B contains only one node unless $d \leq 3$ and $q \equiv \pm 1$ (mod n).*

Proof. Let us first consider the case $q \not\equiv \pm 1$ (mod n). By Lemma 3.4.4, both A and B are consecutive runs. Let q^* be the magnitude of q. Look at edges coming from A. Since at most $d - 1$ of them go to B, there must exist a k-value such that all edges with it coming from A will stay within A. Let k' be such a k-value. That is, for any $i \in A$, $qi + r + k' \in A$. Consider any consecutive pair $i, i + 1 \in A$. Note that $|B| \geq n/2 > q^* - 1$. The $q^* - 1$ nodes between $qi + r + k'$ and $q(i + 1) + r + k'$, $\{qi + r + k' + j \cdot sign(q) \mid j = 1, \ldots, q^* - 1\}$, must belong to A. If A contains x nodes, we find in this way that A has at least $x + (x - 1)(q^* - 1)$ nodes. Thus, $x + (x - 1)(q^* - 1) \leq x$. Since $q^* > 1$, we must have $x = 1$. Now, assume that $q \equiv \pm 1$ (mod n) and $d \geq 4$. Then $n > 4$. By Lemma 3.4.4, A and B are consecutive runs. Suppose to the contrary that $|B| \geq |A| \geq 2$. If A contains only two nodes, then at most 4 of the $2d$ edges emanating from A to A, so that there are at least $2d - 4 \geq d$ edges from A to B, contradicting the assumption on A and B. Thus, A has at least three elements. Write $A = \{a, a + 1, \ldots, a + k\}$ ($k \geq 2$). From the proof of Lemma 3.4.4, we know that there are exactly $2(d - 1)$ edges between A and B among $4d$ edges emerging into nodes $a - 1$, a, $a + k$ or $a + k + 1$. However, since $d \geq 4$, there exists at least one node whose edges go to both $a - 2$ and $a + 1$, which contributes an additional edge between A and B, contradicting the assumption on A and B. $\qquad\qquad\qquad\qquad\qquad\qquad\qquad\qquad\qquad\square$

Lemma 3.4.6 *Suppose that $g > 1$ and that g divides d. If the nodes of $G(d, n, q, r)$ can be partitioned into two disjoint nonempty set A and B such that at most $d - 1$ edges go from A to B, then either A or B has only one element unless*

(i) $G(d, n, q, r)$ is isomorphic to $G(3, 6, 3, 1)$, or

(ii) $g = d = 2$ and $n = ((q)_n + 1)2^s$ for some natural number s.

Proof. Since g divides n, the nodes of $G(d, n, q, r)$ can be partitioned into $n' = n/g$ groups $\bar{i} = \{i, i + n', \ldots, (g - 1)n'\}$. In each group, all nodes have same successors. Let $\bar{G}$ be the digraph with the nodes $\bar{1}, \ldots, \bar{n}'$ such that an edge from $\bar{i}$ to $\bar{j}$ exists iff an edge from i to x for some x in $\bar{j}$ exists in $G(a, n, q, r)$. It is easy to see that $\bar{G}$ is isomorphic to $G(d, n', q, r)$. However, we have to pay attention to the fact that n' may not be bigger than d. Note that $gcd(n', q) \mid g$ and $g \mid d$ imply $gcd(n', d) \mid d$. When $n' > d$, we know that $\bar{G}$ is $(d - 1)$-line-connected (Corollary of Theorem 3.3 in [19]). When $n' \leq d$, we may look at $\bar{G}$ as a multigraph. This multigraph is the union of $G(n', n', q, r), \cdots, G(n', n', q, r + (\ell - 1)n'), G(d - \ell n', n', q, r + \ell n')$ where $\ell = \lfloor d/n' \rfloor$ and $G(n', n', q, r), \cdots, G(n', n', q, r+(\ell-1)n')$ are the complete graph of n' nodes and hence $(n'-1)$-line-connected. Moreover, $G(d-\ell n', n', q, r+\ell n')$ is $(d-\ell n'-1)$-line-connected since $gcd(n', q)|g$ implies $gcd(n', q)|(d-\ell n')$. So, $\bar{G}$ is at least $d - \lceil d/n' \rceil$ $(\geq n' - 1)$ line-connected. Since $n > d \geq g$, we have $n' \geq 2$ and $\lceil d/n' \rceil \leq \min\{g, d-1\}$. So, each node of $\bar{G}$ has at most $\min\{g, d-1\}$ loops.

Define $\bar{A} = \{\bar{i} \mid \bar{i} \subseteq A\}$ and $\bar{B} = \{\bar{i} \mid \bar{i} \subseteq B\}$. Note that each edge of $\bar{G}$ contains g edges of $G(d, n, q, r)$. If $\bar{A} \cup \bar{B}$ contains all nodes of $\bar{G}$, then there are at most $\lfloor (d - 1)/g \rfloor = d/g - 1$ edges from $\bar{A}$ to $\bar{B}$ and $\bar{A}$ and $\bar{B}$ are nonempty. Since $g > 1$, $d/g - 1 \leq d - 2$. Note that if $n' > d$, then $\bar{G}$ is $(d - 1)$-line-connected. Thus, we have $n' \leq d$. So, $\bar{G}$ is at least $(n' - 1)$-connected. It follows that $d/g - 1 \geq n' - 1$, that is, $d \geq n$, contradicting our assumption that $n > d$. Therefore, there exists a node $\bar{x}$ of $\bar{G}$ not in $\bar{A} \cup \bar{B}$, that is, $\bar{x} \cap A \neq \emptyset$ and $\bar{x} \cap B \neq \emptyset$. We claim that there exists only one such node. To see this, consider a node $a \in \bar{x} \cap A$ and a node $b \in \bar{x} \cap B$. Since a and b have the same d successors, there exist d out-edges of a or b going between A and B. If there are two nodes of $\bar{G}$ not in $\bar{A} \cup \bar{B}$, then we will obtain at least $2d$ edges between A and B. If there are two nodes of $\bar{G}$ not in $\bar{A} \cup \bar{B}$, then we will obtain at least $2d$ edges between A and B, contradicting the choice of A and B. Similarly, if $\bar{x} \cap A$ and $\bar{x} \cap B$ both contain at least two nodes, then we can consider another pair of nodes $a' \in (\bar{x} \cap A) \setminus \{a\}$ and $b' \in (\bar{x} \cap B) \setminus \{b\}$ to obtain other d edges between A and B, contradicting the choice of A and B. Therefore, either $\bar{x} \cap A$ or $\bar{x} \cap B$ contains only one element. Without loss of generality, assume that $\bar{x} \cap A$ has only one element. We next show that if $g > 2$ then $\bar{A} = \emptyset$ unless $G(d, n, q, r)$ is isomorphic to $G(3, 6, 3, 1)$.

In fact, if $\bar{A}$ is not empty, then at least $d - \lceil d/n' \rceil$ edges of $\bar{G}$ come to $\bar{A}$ from $\bar{B} \cup \{\bar{x}\}$ since $\bar{G}$ is $(d - \lceil d/n' \rceil)$-line-connected. These edges contains at least $(g - 1)(d - \lceil d/n' \rceil)$ edges of $G(d, n, q, r)$, going from B to A. Since $n > d \geq g$, we have $n' \geq 2$. If $n' > 2$ or d is even, then $d \geq 2\lceil d/n' \rceil \geq \frac{g-1}{g-2}\lceil d/n' \rceil$. Hence,

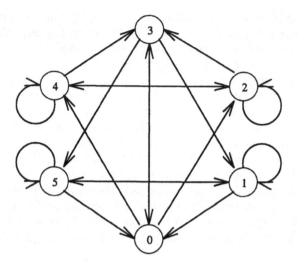

Figure 3.3 $G(3,6,3,1)$ $(A = \{0,1,3,5\}, B = \{2,4\})$

$(d - \lceil d/n' \rceil)(g - 1) \geq d$. It contradicts the hypothesis on A and B. If $n' = 2$ and d is odd, then $|\bar{A}| = 1$, $\bar{B} = \emptyset$ and $n = 2g$. Since $n \geq d$ and $g|d$, we have $d = n$ or g. If $d = n$, then $G(d,n,q,r)$ is the complete digraph of order n. Since $g \geq 3$ and $|A \cap \bar{x}| = 1$, $|B| = g - 1 \geq 2$. Thus, there are at least $2(d - 2)$ edges going from B to A. Because $d = 2g \geq 6$, we have $2(d - 2) > d$, a contradiction. If $g = d$, then note that $\bar{x}$ consists of either all odd nodes or all even nodes. Thus, A receives at least $(d - 1)/2$ edges from each node. If $g > 3$, then B has at least three nodes and hence A receives at least $3(d - 1)/2$ edges from B, contradicting the assumption on A and B. Therefore, $g = d = 3$ and $n = 6$. Thus, $|A| = 4$ and $|B| = 2$. There is only one node in A receiving edges from B. (Otherwise, there are at least four edges from B to A.) Therefore, all in-edges of two nodes in B come from nodes in $\bar{x}$ and all in-edges of two nodes in A which do not receive edges from B come from $A \setminus \bar{x}$. It turns out that $G(d,n,q,r)$ is isomorphic to $G(3,6,3,1)$ as shown in Figure 3.3.

When $g = 2$ and $d > 2$, both $\bar{x} \cap A$ and $\bar{x} \cap B$ have only one element. To show that either A or B has only one element, it suffices to prove that either $\bar{A}$ or $\bar{B}$ is empty. Suppose to the contrary that they both are nonempty. We first claim that no edge exists between $\bar{A}$ and $\bar{B}$. In fact, suppose that there is an edge from $\bar{A}$ to $\bar{B}$. This edge will contributes g edges from A to B. If at least $d - 1$ edges exist from $\bar{A} \cup \{\bar{x}\}$ to $\bar{B}$, then at least $g + (d - 2)(g - 1) = d$ edges exist from A to B, which is impossible. Moreover, $\bar{G}$ is $(d - \lceil d/n' \rceil)$-line-connected and $\lceil d/n' \rceil \leq \lceil n/n' \rceil = g = 2$. So, $\bar{G}$ is $(d - 2)$-line-connected. Thus, there are exactly $d - 2$ edges from $\bar{A} \cup \{\bar{x}\}$ to $\bar{B}$. This implies that $n' < d$ and there are at most $(d - 2)/2 \leq n' - 1$ edges from $\bar{A} \cup \{\bar{x}\}$ to $\bar{B}$ in multigraph $\tilde{G}$ which contains the complete digraph $G(n', n', q, r)$. Hence, $\bar{B}$ must be a singleton. Moreover, since there are exactly $d - 2$ edges from $\bar{A} \cup \{\bar{x}\}$ to $\bar{B}$, there are exactly $g + (g - 1)(d - 3) = d - 1$ edges from A to $B \setminus \bar{x}$. Thus, all in-edges of $B \cap \bar{x}$ come from $B \setminus \bar{x}$ and hence $d = g = 2$, a contradiction. Similarly, it is also impossible that there exists an edge from $\bar{B}$ to $\bar{A}$. Thus, at least $d - 1$ outedges from the node in $A \cap \bar{x}$ go to B and at least $d - 1$ outedges from the node in $B \cap \bar{x}$ go to A. It follows that $\bar{x}$ has at least $d - 1$ edges going to $\bar{A}$ and at least $d - 1$ edges going to $\bar{B}$. Thus, at least $2(d - 1)$ edges go out from $\bar{x}$. It implies $2(d - 1) \leq d$. Hence $d = 2$.

Now, we consider the case $d = 2$ and hence $g = 2$ since $1 < g$ and g divides d. In the above proof, we suppose that $\bar{A}$ and $\bar{B}$ both are nonempty and have proved that no edge exists between $\bar{A}$ and $\bar{B}$ and that $\bar{x}$ has at least $d - 1$ edges going to $\bar{A}$ and at least $d - 1$ edges going to $\bar{B}$. The latter fact implies that $\bar{x}$ has exactly one edge going to $\bar{A}$ and exactly one edge going to $\bar{B}$. Therefore, only one edge goes from $\bar{A} \cup \{\bar{x}\}$ to $\bar{B}$ and only one edge goes from $\bar{B} \cup \{\bar{x}\}$ to $\bar{A}$. First, consider $gcd(q, n') = 1$. If $q \not\equiv \pm 1 \pmod{n'}$, then by Lemma 3.4.5, either $\bar{A} \cup \{\bar{x}\}$ or $\bar{B}$ is a singleton and either $\bar{B} \cup \{\bar{x}\}$ and $\bar{A}$ is a singleton. This happens only if both $\bar{A}$ and $\bar{B}$ are singletons. Thus, $n' = 3$. This is impossible since $q \not\equiv \pm 1 \pmod{n'}$ and $gcd(q, n') = 1$. If $q \equiv 1 \pmod{n'}$, we notice that $\bar{x}$ has no loop, so that $\tilde{G}$ has no loop and is d-line-connected, hence the above $(d - 1)$ size cut cannot exists. If $q \equiv -1 \pmod{n'}$, then $n = 2((q)_n + 1)$, which is the exceptional case. Now, we consider the case that $gcd(q, n') > 1$. Assume by induction that the lemma holds for $G(d, n', q, r)$. Note that $n' = ((q)_n + 1)2^{s'}$ implies $n = ((q)_n + 1)2^{s'+1}$. Thus, if $n \neq ((q)_n + 1)2^s$, then $n' \neq ((q)_n + 1)2^{s'}$. By the induction hypothesis, either $\bar{A} \cup \{\bar{x}\}$ or $\bar{B}$ is a singleton and either $\bar{B} \cup \{\bar{x}\}$ or $\bar{A}$ ia a singleton. It follows that both $\bar{A}$ and $\bar{B}$ are singletons. Hence $n' = 3$. Since $gcd(q, n') > 1$ implies $n' > 3$, we obtain a contradiction. Therefore, either $\bar{A}$ or $\bar{B}$ is empty, so that either A or B is singleton. $\qquad \square$

Lemma 3.4.7 *Suppose that $d \geq 2$ and g divides d. Assume that the nodes of $G(d, n, q, r)$ can be partitioned into two disjoint nonempty sets A and B such that at most $d - 1$ edges go from A to B. Then, both A and B contain loop-nodes.*

Proof. For $g = 1$, if $d \geq 4$ or $q \not\equiv \pm 1 \pmod{n}$ then either A or B contains only one node; such a node must have a loop. By Lemma 3.2.5(a)(b), both A and B have a loop. If $q = 1$, the existence of A and B implies that every node

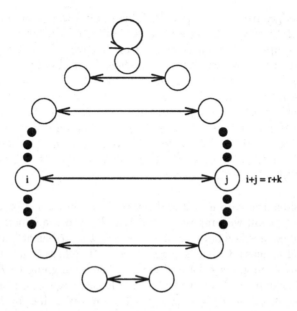

Figure 3.4 $G(1, n, -1, r + k)$

has a loop. For $q = -1$, we note that A and B are consecutive runs. Clearly, there is a k-value such that the edges with this k-value do not go from A to B. As shown in Figure 3.4, all edges with such a k-value provide a symmetric reflection inside of A. If A contains odd number of nodes, then the middle node will be fixed under the reflection and hence get a loop. If A contains even number of nodes, then two consecutive nodes a and $a + 1$ situated at the middle positions of A will reflect to each other. Consider a k-value which is different in one from the former k-value. Such a k-value will give a loop at either node a or node $a + 1$. Thus, A always has a node with a loop. Similarly, we can show that for B.

Now, we consider the case $g > 1$. If either A or B is a singleton then it is obvious. If nether A nor B is singleton, then by Lemma 3.4.6, we have either (1) $G(d, n, q, r)$ is isomorphic to $G(3, 6, 3, 1)$, or (2) $g = d = 2$ and $n = ((q)_n + 1)2^s$ for some natural number s. If (1) occurs, then it is easy to see that A and B both contain loops. If (2) occurs, then let $\bar{A}$, $\bar{B}$ and $\bar{x}$ be the sets and the point in the proof of Lemma 3.4.6. By induction, we may assume that the current lemma holds for $G(d, n', q, r)$. Thus, $\bar{A}$ and $\bar{B}$ both have loops. This imply that A and B both have loops since a node of $\bar{G}$ has a loop iff this node contains a node of $G(d, n, q, r)$ having a loop.   □

3.4.2 Proofs of Theorems 3.4.1-3

Proof of Theore 3.4.1. The first half follows immediately from 3.4.7. To prove the second half, we note that by Lemma 3.2.5(a), each node has at most one loop. It follows that an edge-cut of size less than $d-1$ must separate nodes into two subsets of size at least two. By Lemmas 3.4.5 and 3.4.6, this is possible only in the following cases:

(1) $d \le 3$ and $q \equiv \pi$ (mod n)

(2) $G(d,n,q,r)$ is isomorphic to $G(3,6,3,1)$

(3) $g = d = 2$ and $n = ((q)_n + 1)2^s$ for some natural number s.

These cases can be dealt with by a similar method in the proof of Lemma 3.4.7. □

Proof of Theorem 3.4.2. The first half of the theorem follows from Lemma 3.4.7 and the fact that the modification does not change the indegree at each node. The second half follows from Lemmas 3.4.7, 3.4.5, and 3.4.6. □

Proof of Theorem 3.4.3. If $\lambda = gcd(q - 1, n) = 1$ then by Lemma 3.2.5(e), for every k-value, there is exactly one node with a loop corresponding to the k-value. Such a node is denoted by i_k. Now, we give edges going from i_{k-1} to i_k for $k = 0, \ldots, d - 2$ and from i_{d-1} to i_0 and claim that the resulting digraph has no multiple edge. For otherwise, suppose that the edge from i_{k-1} to i_k has been in $G(d,n,q,r)$. That is, $i_k \equiv qi_{k-1} + r + k'$ (mod n) for some $k' \in \{0, \ldots, d-1\}$. It follows that $i_k \equiv i_{k-1} + k' - k + 1$ (mod n). Multiplying q in both sides and simplifying it, we obtain that $1 + i_k \equiv i_{k-1} + q(k' - k + 1)$ (mod n). Therefore, $1 \equiv (q - 1)(k' - k + 1)$ (mod n). This is impossible since $n > (q' + 1)d$. Similarly, we can show that the edge from i_{d-1} to i_0 is not in $G(d,n,q,r)$. Thus, we have a cyclically-modified $G(d,n,q,r)$ without multiple edges, which, by Theorem 1.1, is d-line-connected.

If $\lambda = gcd(q-1, n) > 1$, then the digraph has either no loop or exactly λ loops for a given k-value. For each loop-node i, $i + n/\lambda$ must also be a loop-node. We edge i to $i + n/\lambda$. This edge is not in $G(d,n,q,r)$ if $q \not\equiv 1$ (mod n). In fact, if the edge is in $G(d,n,q,r)$, then $i + n/\lambda \equiv qi + r + k$ (mod n) for some k in $\{0, \ldots, d-1\}$. That is, $n/\lambda \equiv k - k'$ (mod n) for some $k, k' \in \{0, \ldots, d-1\}$. Thus, $\lambda(k - k') \equiv 0$ (mod n). This is impossible when $q \not\equiv 1$ (mod n). Therefore, when the two exceptional cases stated in Theorem 1.1 does not occur, a modified $G(d, n, q, r)$ exists without multiple edges. By Theorem 1.1, it is d-line-connected.

Now, we consider the exceptional cases. If $q \equiv 1 \pmod{n}$, then either 1 or $n-1$ is not in $\{(r)_n, \ldots, (r+d-1)_n\}$. Thus, we can edge i to either $i+1$ or $i-1$ for all i. This gives a cyclically-modification. If $q \equiv -1 \pmod{n}$ and $d \leq 3$, then n is even and $\lambda = d = 2$ since $\lambda > 1$. In this case, we have exactly two loop nodes with the same k-value. As we proved in the last paragraph, the edges between these two loop nodes are not in $G(d, n, q, r)$. They give a cyclically-modified $G(d, n, q, r)$ which is d-line-connected by Theorem 1.1. Finally, if $g = d = 2$ and $n = ((q)_n + 1)2^s$ then we can prove it by induction on s. For $s = 0$, we have $q \equiv -1 \pmod{n}$. For $s \geq 1$, $G(2, ((q)_n + 1)2^s, q, r)$ is the line digraph of $G(2, ((q)_n + 1)2^{s-1}, q, r)$. Each loop of the former graph is induced by a loop of the latter graph. There is a one-one correspondence between them. The modification for the former graph is inherited from the latter one through this correspondence.   □

3.5   CONNECTIVITY

In this section we show some results of Du, Hsu, and Kleitman [17] on the connectivity of consecutive-d digraphs.

Theorem 3.5.1 *If $g \mid d$ and $g > 1$, then $G(d, n, q, r)$ is $(d - g)$-connected and it is d-connected iff it has no loop.*

Theorem 3.5.2 *Let $g = 1$ and $n > 3d$. Then $G(d, n, q, r)$ is at least $(d - 1)$-connected. Moreover, it is d-connected iff the following does not occur.*

(1) There exists a loop.

(2) $q \equiv -d \pmod{n}$ and $d + 1$ divides n.

(3) $qd \equiv -1 \pmod{n}$ and $d + 1$ divides $q - 1$ and n.

As corollaries, the results of Imase, Soneoka and Okada [32] and Homobono and Peyrat [27] are obtained with improvement on the lower bound of n.

Corollary 3.5.3 (Homobono and Peyrat [27]) $G_I(d, n)$ *is d-connected iff* $g = gcd(n, d) > 1$ *and* $(d+1) \mid n$.

Corollary 3.5.4 (Imase, Soneoka and Okada [32]) *If* $n > d \cdot gcd(n, d)$, *then* $G_B(d, n)$ *and* $G_I(d, n)$ *are at least* $(d-1)$-*connected.*

Corollary 3.5.3 indicates that when $g = 1$ and $(d+1) \mid n$, although $G_I(d, n)$ has no loop, $G_I(d, n)$ is still not d-connected. Thus, in this case, we cannot obtain a d-regular digraph with maximum connectivity by modifying $G_I(d, n)$. However, we can obtain such digraphs by modifying $G_B(d, n)$. The next theorem was obtained by Du, Hsu, and Kleitman [17]. Soneoka, Nakada, and Imase [44] obtained a weaker version independently.

Theorem 3.5.5 *Let* $n > 4d^2$ *and* $d \geq 4$. *Then there is a cyclically-modified* $G_B(d, n)$ *of connectivity d.*

Du, Hsu, and Kleitman [17] furthermore relaxed the condition on n and d by giving the following.

Theorem 3.5.6 *For* $n \geq 2d^2$ *and* $d \geq 2$, *there exists a simply-modified* $G_B(d, n)$ *of connectivity d.*

In the following, we prove the above results.

3.5.1 Consecution Lemma

In this subsection we show a fundamental lemma for studying the connectivity of consecutive-d digraphs.

Recall that a subset of Z_n is a consecutive run if its elements can be consecutively numbered mod n. In a consecutive-d digraph every node has d out-edges ending with a consecutive run of size d. For an easy description, we call the d out-edges from the node a *claw*. Let $g = gcd(n, q)$. Denote $\bar{x} = \{x, x + n/g, \cdots, x + (g-1)n/g\}$. Then all nodes in $\bar{x}$ have the same set of successors. Each $\bar{x}$ will be called an *orbit*. The property of orbits implies that the indegree of a node of $G(d, n, q, r)$ must be divisible by g. Thus, if the indegree of a node is d, then we must have $g \mid d$. It was proved in subsection 3.2.3 that $g \mid d$ iff the indegree of every node is d. Throughout this section, we assume $g \mid d$, i.e., the indegree of every node is d. (To emphasize this, we may still mention this condition in the statements of lemmas and theorems.)

Denote $\hat{x} = \{xg + r, xg + r + 1, \cdots, xg + r + g - 1\}$. Then all nodes in $\hat{x}$ have the same set of predecessors. Each $\hat{x}$ will be called a *block*. Consider a subset S of nodes. A *maximal* consecutive run in S is a subset of S such that no consecutive run in S properly contains it.

Lemma 3.5.7 (Consecution Lemma) *Suppose $g \mid d$ and $g < d$. Let C, D, E be a partition of node set of $G(d, n, q, r)$ such that removal of all nodes in E leaves no path from any node in C to those in D. Let S be the subset of all nodes each of which receives an edge from a node in C. If $|E| < d$, then S is a consecutive run of size at least $|C| + d - g$.*

Proof. Suppose that C intersects y orbits. Each orbit in C provides a consecutive run of size d in S. We call it a C-run. Let x be the number of maximal consecutive runs in S. Note that each maximal consecutive run of size z contains at most $z/g - (d/g - 1)$ different C-runs. Thus,

$$|S|/g - x(d/g - 1) \geq y.$$

Hence, S has at least $g(y + x(d/g - 1))$ elements. Since $S \subseteq C \cup E$, we have $gy + x(d - g) \leq |C| + d - 1$. Note that $gy \geq |C|$. If $g = 1$, then it is clear that $x = 1$. If $d > g > 1$, then $d - g \geq d/2$ since $g \mid d$. Thus, $x = 1$. Finally, $x = 1$ implies that $|S| \geq gy + d - g \geq |C| + d - g$. $\square$

From consecution lemma, it is easy to see that $|E| \geq d - g$. This means that if $g \mid d$ and $g < d$, then $G(d, n, q, r)$ is at least $d - g$ connected.

When $g = d$, $G(d, n, q, r)$ is the line-graph of $G(d, n/d, q, r)$ by Lemma 3.2.8. It was proved in [19] (see Theorem 3.4.1) that if $g \mid d$, then $G(d, n, q, r)$ is at least $(d - 1)$-line-connected and it is d-line-connected iff it has no loop. This implies that if $g = d$ and $n > d^2$, then $G(d, n, q, r)$ is at least $(d - 1)$-connected and it is d-connected iff it has no loop.

3.5.2 Proofs of Theorems 3.5.1-2 and Corollaries 3.5.3-4

Proof of Theorem 3.5.1. The first half of Theorem 3.5.1 has been proved in the last subsection. To prove the second half, suppose that E is a node-cut of the smallest size, which disconnects from C to D, i.e. removal of all the nodes in E leaves no path from nodes in C to those in D. Assume $|E| \leq d - 1$. We will prove the existence of a loop.

Let S be the set of ends of claws from C. By Consecution Lemma, S is a consecutive run of size at least $|C| + d - g$. Without loss of generality, we may assume that all nodes not in S are in D. (Otherwise, we can add them to D without increasing the size of node-cut E.) Since S is a consecutive run, so is its complement D. The following facts are important in the remaining part of the proof.

(a) Every claw from E catches some nodes in D. (Otherwise, E can be decreased, contradicting the minimality of E.)

(b) If an orbit contains an element in C, then it contains no element in E. (Otherwise, E can be decreased by putting such elements into C.) An orbit having an element in C (E) is called a C-orbit (E-orbit).

(c) D has at most $g - 1$ elements in C-orbits. (Otherwise, putting all such elements into C does not change the sets E and S, but makes $|E| + |C| - |S| > g - 1$, contradicting consecution lemma.)

To prove the existence of a loop, we may assume $q^* \geq d$ since a loop always exists for $q^* < d$. We consider two cases.

Case 1. $(g - a)n/g \leq |D| < (g - a + 1)n/g$ for some $a = 0, \cdots, \lfloor g/2 \rfloor$. Each C-orbit contains at least $g - a$ elements in D. Let y be the number of C-orbits. Then D has at least $y(g - a)$ elements in C-orbits. By (c), $y(g - a) \leq g - 1$. So, $y = 1$. It follows that $|S| = d$ and every node in C has a loop.

Case 2. $an/g < |D| \leq (a + 1)n/g$ for some $a = 1, \cdots, \lfloor g/2 \rfloor - 1$. In this case, among g elements in an orbit at least a must be in D and at least $g - a - 1$ must not be in D. Since there are at least $\lceil (|D| + d - 1)/g \rceil$ orbits whose claws hit D, E has at least $(g - a - 1)\lceil (|D| + d - 1)/g \rceil$ elements. So, $d - 1 \geq (g - a - 1)\lceil (|D| + d - 1)/g \rceil$, i.e. $|D| \leq (d - 1)(a + 1)/(g - a - 1) \leq d - 1$. Since $q^* \geq d$, no two claws from adjacent nodes both hit D. Let B be the uppermost block of D. Then B has at most $\lceil g/2 \rceil$ elements whose claws all hit D. Let A be the orbit whose claw hit D in B. When B is removed, only elements whose claws hit $D \setminus B$ have to be moved into E, however, all elements in $A \cap E$ can be removed from E to C. Since $|A \cap E| \geq \lceil g/2 \rceil$, this removal does not increase $|E|$. In this way we can reduce D to have only one block. There are at least d claws hitting D. One of them must from a node in D, which form a loop.

Case 3. $\lfloor g/2 \rfloor n/g \leq |D| < \lceil g/2 \rceil n/g$. This case exists only for g odd and at most two C-orbits exist. If there exists only one orbit, we can prove, as in case 1, that every node in C has a loop. If there are two C-orbits, then each C-orbit must have $(g+1)/2$ elements in C. So, $|E \cup C| \leq d+g$. It follows that the claw of a node in an C-orbit can miss only one block in $E \cup C$. If C has no loop, then $(g+1)/2$ elements of C in another C-orbit must lie in this block. Thus, $1 + ((g+1)/2 - 1)n/g \leq g$. So, $n \leq 2g \leq d$. Thus, $q^* < d$, a loop must exist. $\square$

Proof of Theorem 3.5.2. Suppose there exists a node-cut E of size less than d such that removal the nodes in E leaves no path from C to D. Since $g = 1$, the set S in consecutive lemma is exact the union of C and E. Thus, D is a consecutive run. Note that S contains exactly $|C|$ consecutive runs of size d. They all are end sets of claws from C. It follows that each claw from D or E hits at least one node in D and each claw from C hits at least a node in C. For convenience, we also use the phrase "nodes between two claws", it always means the shorter side between the two claws. We prove the theorem through proving the following two lemmas.

Lemma 3.5.8 *If $|D| \leq |C|$ and $n \geq 3d-2$, then $(q^*-1)(|D|-1) < d$ where q^* is magnitude of q. Furthermore, D has a loop-node unless $q \equiv -d \pmod{n}$.*

Proof. We first show that $C \cup E$ cannot fit between the claws emanating from D. Suppose to the contrary that $C \cup E$ really falls in between some claws from D. Since $C \cup E$ is a consecutive run, it must fall in between two claws from adjacent nodes in D. The number of nodes between two claws from adjacent nodes is $d + q^*$. Since each claw from D must catch a node in D, we have $|C \cup E| \leq d + q^* - 2$. Thus, $q^* > |C|$. Moreover, $n \geq 3d - 2$ and $|C| \geq |D|$ imply $q^* > |C| \geq (n - d + 1)/2 \geq n/3$. Thus, there are at least $\lfloor (|D| - 1)/2 \rfloor$ consecutive pairs in D such that $C \cup E$ are between the claws from each pair. This can happen only if $q^* + d - 2 - \lfloor (|D| - 1)/2 \rfloor \geq n - |D|$ since any two claws end at different groups of nodes. Thus, $q^* + d - 2 \geq n - |D| + |D|/2 - 1$, that is, $q^* \geq n - (d-1) - \|D|/2 > n/2$, a contradiction.

Since $C \cup E$ cannot fit between two claws emanating from two consecutive nodes in D, all nodes between them must be in D. Counting in this way, D should have at least $2 + q^* - d + (|D| - 2)q^*$ nodes, which is less than $|D|$. Therefore, $(q^* - 1)(|D| - 1) < d$.

If $|D| = 1$, then the node in D is clearly a loop-node. If $|D| \geq 2$ and $q^* > 1$, then $q^* - 1 < d$. When $q^* < d$, every two claws from adjacent nodes overlap each other. By Lemma 3.5, D has a loop-node. When $q^* = d$, D has either a loop-node or a pair of nodes x and $x + 1$ such that claws from x and $x + 1$ end with $f(x) = \{x + 1, \cdots, x + d\}$ and $f(x + 1) = \{x - d + 1, \cdots, x\}$, respectively. The latter one implies that $r + qx \equiv x + 1 \pmod{n}$ and $r + q(x + 1) \equiv x - d + 1 \pmod{n}$. Thus, $q \equiv -d \pmod{n}$, the exceptional case. $\qquad \square$

Let m be the multiplicative inverse of $q \pmod{n}$. Let m^* be the magnitude of m.

Lemma 3.5.9 *If $|C| \leq |D|$ and $n > 3d$ then $(m^* - 1)(|C| - 1) < d$. Furthermore, C has a loop-node unless $qd \equiv -1 \pmod{n}$.*

Proof. Denote $c = |C|$. Since $g = 1$, the claws coming out from C can be ordered so that the second edge of each is the first edge of the next claw. Then, C must consist of nodes of index $a, a + m^*, \ldots, a + (c - 1)m^*$, and these must all lie among the $d - 1 + c$ consecutive nodes in $C \cup E$. If $m^* = 1$ the lemma holds trivially. If $m^* > 1$, then either all the nodes in between these nodes of C are in E, so that $(m^* - 1)(c - 1) < d$, or the size of D, $|D|$, is at most $m^* - 1$ so that D can fit between adjacent nodes of C in this order. The lemma is proven if we show that this latter contingency cannot happen when $n > 3d$.

Under the given circumstances we must also have $|D| + c + d - 1 = n$, $n > 3d$, $m^* > |D| \geq c$, $m^* < n/2$ and either $m^* \geq |D| + \lfloor c/2 \rfloor$ (if every other interval of m^* contains D) or $n - 2m^* \geq |D|$ (if some interval do and some do not). If the former occurs then $n/2 \geq |D| + c/2 \geq (3/2)c$, so that $n \leq |D| + c + d - 1 < n/2 + n/6 + d/3 - 1 < n$, a contradiction. If the latter case occur, then $n \geq 2m^* + |D| \geq 3|D| \geq 3c$, so $n \leq |D| + c + d - 1 \leq n - 1$, again, a contradiction.

If $|C| = 1$, then the node in C is obviously a loop-node. If $|C| \geq 2$, then $m^* - 1 < d$. When $m^* < d$, by Lemma 3.5 C contains a loop-node. When $m^* = d$, by Lemma 2.9 C has either a loop or a pair of nodes x and $x + m^*$ such that the claws from x and $x + m^*$ end with $\{x + 1, \cdots, x + d\}$ and $\{x, \cdots, x + d - 1\}$, respectively. The latter one implies $r+qx \equiv x+1 \pmod{n}$ and $r+q(x+m^*) \equiv x \pmod{n}$. Thus, $qd = qm^* \equiv -1 \pmod{n}$, the exceptional case. □

Finally, we complete the proof of Theorem 4.2 by noting that when $q \equiv -d \pmod{n}$, $G(d, n, -d, r)$ has no loop only if $(d+1) \mid n$ and when $qd \equiv -1 \pmod{n}$, the graph has no loop only if $(d+1) \mid (q-1)$ and $(d+1) \mid n$. □

Proof of Corollary 3.5.3. By Lemma 3.2.5(c), $G_I(d, n)$ has no loop iff $(d+1) \mid n$. Thus, by Theorem 3.5.1, if $g > 1$ and $(d+1) \mid n$, then $G_I(d, n)$ is d-connected. If $g = 1$ and $(d+1) \mid n$, then by Theorem 3.5.2, $G_I(d, n)$ is not d-connected. Therefore, $G_I(d, n)$ is d-connected iff $g > 1$ and $(d+1) \mid n$. □

Proof of Corollary 3.5.4. By Theorem 3.5.2, we may assume that $1 < g = gcd(n, d) < d$. Consider the proof of Theorem 3.5.1. We show $|E| \geq d - 1$. In case 1, if $|C| = 1$, $|E| \geq d - 1$; if $|C| \geq 2$, then we must have $1 + n/g \leq d$ since only one C-orbit exists. Thus, $n \leq g(d-1)$, a contradiction. In case 2, by the reduction, we may assume that D has only one block. Since $n > gd \geq (g-1)d$, D has exactly one loop. However, there are d claws hitting D. $d-1$ of them must come from E. That is, $|E| \geq d-1$. In case 3, if there is only one C-orbit, then it is similar to that in case 1. If two C-orbits exist, then each C-orbit contains $(g-1)/2 \; (\geq 2)$ elements of C. Between two such elements of distance n/g exactly two are in C and the rest are in E. So, if $|E| \leq d-2$, then $n/g \leq d$, a contradiction. (We remark that in the latter subcase, $|D| < 2|E|+g+1 < 3d$. This remark will be used later.) □

3.5.3 Proof of Theorems 3.5.5-6

We first prove a lemma.

Lemma 3.5.10 *Let* $n > \max(5d, d \cdot gcd(n, d))$ *and* $d \geq 2$. *Suppose in* $G_B(d, n)$ *that* E *is a node-cut of size at most* $d - 1$ *such that removal the nodes in* E *leaves no path from* C *to* D. *Then either*

(1) $|C| \leq d$ *and* C *has a loop, or*

(2) $|D| \leq d$ *and* D *has a loop.*

Furthermore, a loop-node in E *is within* $d - 1$ *from* C *and also within* $d - 1$ *from* D.

Proof. First, assume $gcd(n, d) = 1$. Then (1) or (2) follows from 3.5.8 or 3.5.9 in the proof of Theorem 3.5.2. By consecution lemma, the set S has $|C| + d - 1$ elements. Therefore, $C \cup E (= S)$ is a consecutive run and every node in E has a successor in D. These two facts imply respectively that every loop-node in E is within distance $d - 1$ from C and within distance $d - 1$ from D.

Now, we assume $1 < gcd(n, d) < d$. We notice that in the proof of Theorem 3.5.1, the minimality of $|E|$ is assumed. Here, we did not assume it. However, by Corollary 3.5.4, $|E| = d - 1$ is indeed minimum. A little difference is that D may not consecutive. To meet the assumption $S = C \cup E$ in the proof of Theorem 3.5.1, we have to move at most $g - 1$ elements from C to D. Those elements are in $(C \cup E) \setminus S$ and cannot have a loop. So, the movement affects only the sizes of C and D. Suppose C' and D' are obtained from C and D respectively through the movement. Now, consider the proof of Theorem 3.5.1 applying to E, C' and D'. In case 1, $|C'| \leq g$ and every node in C' has a loop, so that $|C| \leq 2g - 1 \leq d - 1$. In case 2, $|D'| \leq d - 1$ and D' has a loop, so that $|D| \leq d - 1$ and D has a loop. In case 3, if only one C'-orbit exists, then it is the same as in case 1; if two C'-orbits exist, then $|D'| < 3d$, so that $n < 5d$. Next, we consider a loop-node x in E. x's claw must hit D since, otherwise, x can be removed from E, contradicting the minimality of E. This implies that x is within distance $d - 1$ from D. Since $C' \cup E$ is consecutive, x is also within distance $d - 1$ from C' and hence from C.

Finally, we consider the case of $gcd(n, d) = d$. Note that $G_B(d, n)$ is the line-graph of $G_B(d, n/d)$. Thus, E gives a line-cut of size at most $d - 1$ for $G_B(d, n/d)$. However, we will prove in the next section that such a line-cut must be incident to a node of $G_B(d, n/d)$, which implies that C or D is a singleton. So, the lemma holds. □

Proof of Theorem 3.5.5. Consider two loop-nodes x and y with distance at least $2d - 1$. When (1) in Lemma 3.5.10 occurs, $x \in C$ will implies $y \in D$. When (2) in Lemma 3.5.10 occurs, $x \in D$ implies $y \in C$. This means that as long as all loop-nodes are connected by a cycle (or disjoint cycles) with edges of distance at least $2d - 1$, the node-cut E of size less than d will no longer exist in the modified graph. Hence, the connectivity becomes d. We next show the existence of such a modification. Consider a graph H with node set consisting of all loop-nodes of $G_B(d, n)$ and an edge between x and y exists iff x and y are apart from distance at least $2d - 1$. If H is Hamiltonian, then the theorem is proved. We prove the hamiltonian property of H by showing that degree of H is bigger than half the number of its nodes. Consider any loop-node x of $G_B(d, n)$. Let A be the set of nodes in $G_B(d, n)$, within $2d - 1$ from x and B the set of nodes not in A. Note that $n > 12d$ and $4d < n/(d-1) \leq n/3$. By moving

A in two ways with distance $n/(d-1)$ we can obtain two disjoint 'copies' of A. By Lemma 3.4.7, for each loop-node x in A there is a corresponding loop-node in each copy. If two loop-nodes in A are not adjacent, then the corresponding loop-nodes in the copy are distinct. So, each copy contains loop-nodes at least half the number of loop-nodes in A. It follows that B has loop-nodes more than A has. This completes the proof.   □

Proof of Theorem 3.5.6. Consider two cases.

Case 1. $\gcd(n, d-1) = 1$. In this case, there exists a unique loop for each k-value. Let i_k denote such a loop-node. Then $(d-1)(i_k - i_{k-1}) + 1 \equiv 0$ (mod n). Hence, $|i_k - i_{k-1}| \geq (n-1)/(d-1) > 2d - 1$. Connecting by loop-nodes by edges from i_{k-1} to i_k ($i_d = i_0$). We obtain a cycle meeting the requirement in the proof of Theorem 3.5.5.

Case 2. $\gcd(n, d-1) = \lambda > 1$. For each k-value there are exactly λ loop-nodes which are evenly distributed with distance n/λ. Note that $n/\lambda \geq 2n/d \geq 4d$. We connect each loop-node x to another loop-node $x + n/\lambda$. Then all loop-nodes are connected by several disjoint cycles of size λ with all edges in the graph H in the proof of Theorem 3.5.5.

Finally, we notice that the above connections give no multiple edge. The detail can be found in the proof of Theorem 3.4.3 in the next section.   □

3.6   SUPER LINE-CONNECTIVITY

A digraph is said to have *super line-connectivity* if its line-connectivity equals the minimum degree (outdegree and indegree) and every minimum edge-cut consists of edges incident to the same node. A digraph having super line-connectivity reaches the maximum reliability in certain sense [43].

Soneoka [43] proved that if $n \geq d^3$ and $d \geq 3$, then cyclically-modified $G_B(d, n)$ has super line-connectivity. A corresponding result for consecutive-d digraphs was obtained by Cao, Du, Hsu, Hwang, and Wu [9] as follows.

Theorem 3.6.1 *Suppose g divides d. Then for $d \geq 5$, every modified $G(d, n, q, r)$ has super line-connectivity and for $d \geq 3$, every cyclically-modified $G(d, n, q, r)$ has super line-connectivity unless*

(1) $g = 1$, $d = 3$ and $q = \pm 1$, or

(2) $G(d, n, q, r)$ is isomorphic to $G(3, 6, 3, 1)$.

Corollary 3.6.2 (Soneoka [43]) *If $n > d \geq 3$, then every cyclically-modified $G_B(d, n)$ has super line-connectivity.*

The proof of this theorem uses the same techniques introduced in previous two sections. We sketch it as follows.

First, recall that we already proved the following.

Suppose g divides d and $d \geq 3$. If the nodes of $G(d, n, q, r)$ can be partitioned into two disjoint nonempty set A and B such that at most $d - 1$ edges go from A to B, then either A or B has only one element unless

(1) $g = 1$, $d = 3$ and $q = \pm 1$, or

(2) $G(d, n, q, r)$ is isomorphic to $G(3, 6, 3, 1)$.

Consider a minimum edge-cut C in a modified $G(d, n, q, r)$. Clearly, its size is d. If it contains an edge not in $G(d, n, q, r)$, then the edge-cut induces an edge-cut C' of size at most $d - 1$ in $G(d, n, q, r)$. From the above fact, C' isolates a node in $G(d, n, q, r)$ unless (1) or (2) occurs. Thus, if (1) and (2) do not occurs, then the original edge-cut C has to isolate a node in the modified $G(d, n, q, r)$.

Next, we study the case that all edges in C belong to $G(d, n, q, r)$. Clearly, it suffices to prove the following.

Theorem 3.6.3 *Assume $d \geq 3$ and $g \mid d$. Let C be an edge-cut of size d in a modified $G(d, n, q, r)$ such that all edges in C belong to $G(d, n, q, r)$. Let A and B form a partition of the node set of $G(d, n, q, r)$. Suppose that removal C leaves no path from A to B. Then either A or B contains only one element unless $d \leq 4$ and the modification is not cyclic.*

The proof is then divided into the following lemmas.

Lemma 3.6.4 *Let $d \geq 3$ and $g \mid d$. Then $|A| \neq 2$ unless $d \leq 4$ and the modification is not cyclic.*

Lemma 3.6.5 *Let $g = 1$ and $d \geq 3$. Then both A and B are consecutive runs unless $d \leq 4$ and the modification is not cyclic.*

Lemma 3.6.6 *If $g = 1$ and $d \geq 3$, then either A or B contains only one node unless $d \leq 4$ and the modification is not cyclic.*

Lemma 3.6.7 *If $g > 1$, $d \geq 3$, and g divides d, then either A or B has only one element unless $d \leq 4$ and the modification is not cyclic.*

3.7 HAMILTONIAN PROPERTY

The Hamiltonian property is an additional nice thing for consecutive-d digraphs to have. Define $g = gcd(n, q)$ and $n' = n/g$. In this section, we study the Hamiltonian property of consecutive-d digraphs. The following two theorems were obtained by Du, Hsu and Hwang [12].

Theorem 3.7.1 *If $d < g$ then $G(d, n, q, r)$ is not Hamiltonian.*

Proof. Patition the n vertices of $G(d, n, q, r)$ into n' groups of g vertices where the group g_i consists of vertices $\{i, i + n', ..., i + (g - 1)n'\}$. Then vertices in the same group have the same set of d successors. Therefore there are at most $n'd < n'g = n$ successors, i.e., some vertices have no in-edge. Thus, $G(d, n, q, r)$ cannot be Hamiltonian. $\square$

Theorem 3.7.2 *If $1 < g \leq d$ then $G(d, n, q, r)$ is Hamiltonian.*

Proof. Partition the n vertices into n' groups as in the proof of the last theorem. Define a digraph G' with $g_1, ..., g_{n'}$ as vertices and each g_i has g out-edge going to the g groups containing the g successors of vertices in g_i (with respect to $G(d, n, q, r)$). Label these out-edges by the corresponding successors. Then the $n'g$ edges in G' carry distinct labels. This implies that G' is regular. By the construction of G' it is easy to verify that $G(g, n, q, r)$ is the line digraph of G'. Furthermore, G' is in fact the digraph $G(g, n', q, r')$ where $r' \equiv r \pmod{n'}$. If $gcd(n', q) = 1$, then in the subgraph $G(2, n', q, r')$ every pair of vertices g_i, g_{i+1} are adjacent to a distinct vertex. Hence $G(2, n', q, r')$ is connected. If

$gcd(n', q) > 1$, then by induction on n', $G(g, n', q, r')$ is Hamiltonian, hence connected. In any case, we have shown that $G(g, n', q, r')$ is connected and regular, hence it is eulerian. Therefore, $G(g, n, q, r)$ is Hamiltonian.   □

In the following, we study the case $g = gcd(n, q) = 1$. First, we give an account of the general approach.

3.7.1   A General Approach

The first step is to select n edges of $G(d, n.q, r)$ which form a 1-factor F, i.e., a subgraph such that every vertex i, $0 \leq i \leq n - 1$, has one indegree and one outdegree. Let $C_1, ..., C_m$ be the set of disjoint circuits of F. If $m = 1$, then F is a Hamiltonian circuit; hence $G(d.n.q.r)$ is Hamiltonian. If $m > 1$, then we want to merge the m circuits into a single circuit. This is done by merging two circuits at a time. Suppose that vertices i and j lie on two different circuits C_x and C_y. Let i' (j') be the vertex preceding i (j) on C_x (C_y). Then we can merge C_x and C_y by replacing the two edges $i' \rightarrow i$ and $j' \rightarrow j$ by the two edges $i' \rightarrow j$ and $j' \rightarrow i$. We call such a replacement the interchange of (i, j) and we say i and j are *interchangeable* if $i' \rightarrow j$ and $j' \rightarrow i$ are edges of $G(d, n, q, r)$. If after the merging of C_x and C_y the number of circuits is still greater than one, then we repeat the same procedure until only one circuit is left.

Note that i and j may be interchangeable at the beginning but not after some other interchanges involving i or j have taken place. For example, $G(3.8, 3, 0)$ contains edge $0 \rightarrow 6$, $0 \rightarrow 5$, $3 \rightarrow 6$ and $3 \rightarrow 5$. So if F contains the two edges $0 \rightarrow 6$ and $3 \rightarrow 5$ on C_x and C_y, respectively, vertices 5 and 6 are interchageable. But, if the interchange $(4, 5)$, which replaces edges $3 \rightarrow 5$ and $1 \rightarrow 4$ by the two edges $3 \rightarrow 4$ and $1 \rightarrow 5$, has taken place, then vertices 5 and 6 are no longer interchangeable since the replacement would areate the edge $1 \rightarrow 6$ which is not in $G(3, 8, 3, 0)$. Therefore, a set of interchanges can be given only when their relative ordering is specified.

Let S denote a set of edges in $G(d, n, q, r)$ such that the undirected version of $S \cup F$ is connected. Then there exists a set of $m - 1$ edges $S' \subseteq S$ such that the undirected version of $S' \cup F$ is also connected. If we can show that there exists an ordering R such that each edge in S' represents an interchageable pair under R, then we can merge the m circuits into one by making the interchanges specified in S'.

Therefore, to check whether $G(d, n, q, r)$ is Hamiltonian, it suffices to do the following three things:

(i) Find a 1-factor F of $G(d, n, q, r)$.

(ii) Find a set of edges S such that the undirected version of $S \cup F$ is connected.

(iii) Find an ordering R such that each edge in S' represents an interchangeable pair under R.

3.7.2 $d \geq 5$

Theorem 3.7.3 *If $g = 1$ and $d \geq 5$ then $G(d, n, q, r)$ is Hamiltonian.*

Proof. Let F consist of the edges $i \rightarrow qi + r + 2, i = 0, 1, ..., n - 1$. Since $gcd(n, q) = 1$, it is easy to see that F is a 1-factor. Suppose that F consists of m circuits with $m > 1$. Let S consist of the edges $i \rightarrow i+1, i = 0, 1, ..., n-2$. Then clearly, the undirected version of $S \cup F$ is connected since S itself is connected. Let R be an ordering such that any interchange of the type $(2i, 2i + 1)$ precedes all interchanges of the type $(2i - 1, 2i)$ in S'. We now verify that each edge in S' represents an interchangeable pair under R.

Under R, we do all interchanges of type $(2i, 2i+1)$ in S' before the interchanges of type $(2i - 1, 2i)$ (the ordering of interchanges of the same type is not important). Note that the effect of the interchanges of the first type is to replace an edge $i' \rightarrow i$ by either $i' \rightarrow i - 1$ or $i' \rightarrow i + 1$. Meanwhile, the effect of the interchanges of the second type is to replace an edge $i' \rightarrow j$ to either $i' \rightarrow j - 1$ or $i' \rightarrow j + 1$ where $j \in \{i - 1, i, i + 1\}$. So, the cumulative effect of the interchanges in S' under R is to replace $i' \rightarrow i$ by $i' \rightarrow i \pm 2$. Since $i = qi' + r + 2$ and $d \geq 5$, the edge $i' \rightarrow i \pm 2$ remains an edge of $G(d, n, q, r)$. The theorem is proved.   □

3.7.3 $d \leq 4$

It was also conjectured by Du, Hsu and Hwang [12] that for $d \geq 3$ and $g = 1$, every generalized de Bruijn-Kautz digraphs is Hamiltonian. However, so far, only partial solution is obtained. Du, Hsu, Hwang and Zhang [13] showed that for $d \geq 3$, the generalized de Bruijn digraphs and the generalized Kautz digraphs are Hamiltonian. Du and Hsu [14] extended the result by proving the following theorem.

Theorem 3.7.4 (Du and Hsu [14]) *If* $g = 1$, $q \leq d$ *and* $3 \leq d$ *then* $G(d, n, q, r)$ *is Hamiltonian.*

For $d = 1, 2$, there are the following results.

Theorem 3.7.5 (Du and Hsu [14]) $G(2, n, q, r)$ *is Hamiltonian iff either* $G(1, n, q, r)$ *or* $G(1, n, q, r + 1)$ *is Hamiltonian.*

Proof. The "if" part is trivial. We only show the "only if" part. For any digraph G with vertex set V and edge set E, let G^* denote the bipartite graph with vertex set $V \cup V^*$ where $V^* = \{v^* \mid v \in V\}$, and with edges from each vertex u to vertex v if edge $(u, v) \in E$. Let ϕ denote the map from edge (u, v) of G to edge (u, v^*) of G^*. Obviously, ϕ is one-to-one and onto. It is a useful fact that a subset H of E forms a 1-factor of G iff $\phi(H)$ forms a perfect matching of G^*. To show the "only if" part, it suffices to prove that $G^*(2, n, q, r)$ is a cycle using edges of $G^*(1, n, q, r)$ and $G^*(1, n, q, r + 1)$, alternatively. Since $gcd(n, q) = 1$, for every vertex j, we can find a vertex i such that $j \equiv qi + r \pmod{n}$, that is, (i, j^*) and $(i, (j + 1)^*)$ are two edges of $G^*(2, n, q, r)$. It follows that $G^*(2, n, q, r)$ is a cycle passing through $0^*, 1^*, \ldots, (n - 1)^*$ consecutively and edges of $G^*(1, n, q, r)$ and $G^*(1, n, q, r + 1)$ alternatively. $\square$

Theorem 3.7.6 (Hwang[28]) *Suppose that* $q = \prod_{i=1}^{t} p_i^{a_i}$. *Then* $G(1, n, q, r)$ *is Hamiltonian iff* $n = \prod_{i=1}^{t} p_i^{b_i}$ *where* $b_i = 0$ *if* p_i *divides* g *and* $d_i \leq 1$ *if* $p_i^{a_i} = 2$ *not dividing* g.

3.7.4   Cycle-Triangle Conjecture

In 1986, Du, Hsu, and Hwang [12] made the following conjecture.

Conjecture 1 (Cycle-Triangle Conjecture) *Let* F *be the unioun of* n *disjoint triangles and a cycle of the* $3n$ *vertices of the* n *triangles. Then the stability number (the cardinality of the largest independent set of vertices) is* n.

The initial motivation for this conjecture is to establish Hamiltonian property of consecutive-d digraphs, With this conjecture, they was able to prove that for $d \geq 7$, $G(d, n, q, r)$ is hamiltonian. Later, Du, Hsu, and Hwang found a better way to go around this conjecture. But, this conjecture has already stood in the literature independently from the original motivation. In 1987, D. Frank Hsu spread the cycle-triangle conjecture in a conference in Florida. Paul Erdös was interested in it and made a stronger version.[47]

Conjecture 2 *Let F be the unioun of n disjoint triangles and a cycle of the $3n$ vertices of the n triangles. Then F is three vertex-colorable, i.e, $\chi(F) = 3$.*

The popularity of the conjecture was due to Erdös. As Erdö travelled around the world, the conjecture was promoted every where. M.R. Fellow [22] and N. Alon and M. Tarsi [1] made some progress and the conjecture was finally proved by H. Fleischner and M. Stiebitz [24] in 1992. This seems an end of the story. However, it is not. Recently, we found an interesting continuation.

To study broadcasting in some nonblocking switch networks, Hwang [29] made the following conjecture.

Conjecture 3 (Hwang [29]) *Consider a graph with colored edges. If each color and each vertex has at most d edges, then the edge set can be partitioned into $2d$ or less matchings each of which contains at most one edge of a color.*

This conjecture is equivalent to the following.

Conjecture 4 *Consider the line graph G of a d-regular graph. Partition the vertex set of G into disjoint subsets of size at most d and for each subset, construct a clique on it. Then the resulting graph is $2d$ vertex-colorable.*

This conjecture has been verified for $d = 2$ and $d = 3$ [29, 25]. Clearly, conjecture 4 is not a natural generalization of conjecture 3. By combining them, we propose the following.

Conjecture 5 *Consider the line graph G of a d-regular graph. Partition the vertex set of G into disjoint subsets of size exactly n (at most n) with $d \leq n \leq 2d - 1$ and for each subset, construct a clique on it. Then the resulting graph G^* is $2d - 1$ ($2d$) vertex-colorable, i.e., $\chi(G^*) = 2d - 1$ ($\chi(G^*) = 2d$).*

This provides an interesting open problem.

3.7.5 de Bruijn-Good Graphs

There is an interesting application of Hamiltonian cycle of de Bruijn digraph in coding theory. A Hamiltonian cycle in de Bruijn digraph corresponds to an M-sequence in coding theory. To construct Hamiltonian cycle in $B(2,t)$, one can also use a homomorphism from $B(2,t)$ to $B(2,t-1)$ which maps $(a_1, a_2, \cdots, a_t)$ to $(a_1 + a_2, a_2 + a_3, \cdots, a_{t-1} + a_t)$ where the operation is in $GF(2)$. (See A. Lempel [36] for detail.) Motivated from this application, Wan and Liu [46] determined all homomorphisms from $B(2,t)$ to $B(2,t-1)$.

Theorem 3.7.7 $B(2,t)$ *has only two graph automorphisms, the identity and the dual. (The dual automorphism exchanges 0 and 1.) There are only six homomorphisms from $B(2,t)$ to $B(2,t-1)$ denoted by*

$$\mathcal{D}, \mathcal{D}^*, \mathcal{D}', \mathcal{DD}, \mathcal{DD}^*, \mathcal{DD}'$$

where D is the dual automorphism and

$\mathcal{D}: (a_1, a_2, \cdots, a_t) \to (a_1 + a_2, a_2 + a_3, \cdots, a_{t-1} + a_t)$.

$\mathcal{D}^*: (a_1, a_2, \cdots, a_t) \to (a_2, a_3, \cdots, a_t)$.

$\mathcal{D}': (a_1, a_2, \cdots, a_t) \to (a_1, a_2, \cdots, a_{t-1})$.

Recently, one is interested in 2-dimensional arrays [21]. Let m, n be positive integers. Let M be an arbitrary set. The de Bruijn-Good graph over M with dimension 2 and order (m, n) is the digraph with vertex set V consisting of all $m \times n$ matrices $(\alpha_{ij})_{m \times n}$ over M, i.e.,

$$V = \{(\alpha_{ij})_{m \times n} \mid \alpha_{ij} \in M, 1 \leq i \leq m, 1 \leq j \leq n\},$$

and edge set consisting of all edges in the forms

$$(\alpha_{ij})_{m \times n} \to (\alpha_{i+1,j})_{m \times n}$$

and

$$(\alpha_{ij})_{m \times n} \to (\alpha_{i,j+1})_{m \times n}.$$

Liu et al [37, 38, 39] determined homomorphisms and automorphisms of de Bruijn-Good graphs. However, graph properties of 2-D de Bruijn-Good graphs have not been well-studied.

Acknowledgement

Subsection 3.7.5 is based on materials and letters provided by Professor Mulan Liu. Authors wish to thank her for her kind and insightful suggestions.

REFERENCES

[1] N. Alon and M. Tarsi, Colorings and Orientations of graphs, *Combinatorica* to appear.

[2] J.C. Bermond, N. Homobonpo, and C. Peyrat, Large fault-tolerant inter-connection networks, *Graphs and Combinatorics*, (1989).

[3] J.C. Bermond, F. Comellas, and F. Hsu, Distributed loop computer net-works: a survey, *IEEE Trans. on Computers*, to appear.

[4] J.C. Bermond and C. Peyrat, de Bruijn and Kautz networks: a competitor for the hypercube? preprint.

[5] J.C. Bermond and C. Peyrat, Broadcasting in de Bruijn networks, Proc. 19-th Southeastern Conference on Combinatorics, graph Theory and Com-puting, 1988.

[6] J.C. Bermond, N. Homobono, and C. Peyrat, Connectivity of Kautz net-works, *French-Israel Conference on Combinatorics*, Jerusalem, 1988.

[7] J. Bond and A. Ivanyi, Modeling of interconnection networks using de Bruijn graphs, *Proc. Thied Conference of Program Designers* (A. Vitanyi ed.), Budapest, (1987) 75-88.

[8] N.G. de Bruijn, A combinatorial problem, *Nederl. Akad. Wetensh. Proc.* 49 (1946) 758-764.

[9] F. Cao, D.-Z. Du, D.F. Hsu, L. Hwang, and W. Wu, Super-line-connectivity of consecutive-*d* digraphs, Technical Report, Department of Computer Sci-ence, University of Minnesota, 1995.

[10] F.K. Chung, Diameter of communication networks, *Proceedings of Sym-posia in Applied Mathematics*, 34 (1986) 1-18.

[11] D.-Z. Du, D.F. Hsu, and F.K. Hwang, Doubly-linked ring networks, *IEEE Trans. on Computers*, C-34 (1985) 853-855.

[12]D.-Z. Du, D.F. Hsu, and F.K. Hwang, Hamiltonian property of d-consecutive digraphs, *Mathematical and Comput. Modeling* 17: 11 (1993) 61-63.

[13]D.-Z. Du, D.F. Hsu, F.K. Hwang and X.M. Zhang, The Hamiltonian property of generalized de Bruijn digraphs, *J. Comb. Theory (Series B)*, 52 (1991) 1-8.

[14]D.-Z. Du and D.F. Hsu, On Hamiltonian consecutive-d digraphs, *Banach Center Publication*, 22 (1989).

[15]D.-Z. Du and F.K. Hwang, Generalized de Bruijn digraphs, *Networks*, 18(1988) 27-33.

[16] D.-Z. Du, D.F. Hsu, and Y.D. Lyuu, On the diameter vulnerability of Kautz digraphs, to appear.

[17] D.-Z. Du, D.F. Hsu, and D.J. Kleitman, On connectivity of consecutive-d digraphs, manuscript.

[18] D.-Z. Du, D.F. Hsu, and D.J. Kleitman, Modification of consecutive-d digraphs, to appear.

[19]D.Z. Du, D.F. Hsu and G.W. Peck, Connectivity of consecutive-d digraphs, *Discrete Applied Mathematics* 37/38 (1992) 169-177.

[20]J. Fabrega, M.A. Fiol, and M. Escudero, The connectivity of iterated line digraphs, *J. of Graph Theory*, to appear.

[21] C.T. Fan, S.M. Fan, S.L. Ma, and M.K. Siu, On de Bruijn arrays, *Ars Combin.* 19A (1985) 205-213.

[22] M.R. Fellow, Transversals of vertex partitions of graphs, *SIAM J. Disc. Math.* 3 (1990) 206-215.

[23]M.A. Fiol, J.L.A. Yebra, and I. Alegre, Line digraph iterations and the (d, k) digraph problem, *IEEE Trans. on Computers*, C-33 (1984) 702-713.

[24] H. Fleischner and M. Stiebitz, A solution to a colouring problem of P. Edös, manuscript.

[25] B. Gao, private communication.

[26]N. Homobono, Connectivity of generalized de Bruijn and Kautz graphs, *Proc. 11-th British Conference, Ars Combinatoria*, (1988).

[27]N. Homobono and C. Peyrat, Connectivity of Imase and Itoh digraphs, *IEEE Trans. on Computers*, (1988).

[28]F.K. Hwang, The Hamiltonian property of linear functions, *Operations Research Letters*, 6 (1987) 125-127.

[29] F.K. Hwang, Broadcasting in a 3-stage point-to-point nonblocking networks, manuscript.

[30]M. Imase and M. Itoh, Design to minimize a diameter on Building block network, *IEEE Trans. on Computers*, C-30 (1981) 439-443.

[31]M. Imase and M. Itoh, A design for directed graph with minimum diameter, *IEEE Trans. on Computers*, C-32 (1983) 782-784.

[32]M. Imase, T. Soneoka, and K. Okada, Connectivity of regular directed graphs with small diameter, *IEEE Trans. on Computers*, C-34 (1985) 267-273.

[33]M. Imase, I. Soneoka, and K. Okada, A fault tolerant processor interconnection network, *Systems and Computers in Japan*, 17:8 (1986) 21-30.

[34]W.H. Kautz, Bounds on directed (d, k) graphs, *Theory of cellular logic networks and machines, AFCRL-68-0668 Final Report*, (1968) 20-28.

[35]V.P. Kumar and S.M. Reddy, A class of graphs for fault-tolerant processor interconnections, *IEEE 1984 Int. Conf. Distributed Computing Systems*, (1984) 448-460.

[36] A. Lempel, On a homomorphism of the de Bruijn graph and its applications to the design of feedback shift registers, *IEEE Trans. Computers* C-19 (1970) 1204-1209.

[37] M.L. Liu, Homomorphisms and automorphisms of 2-D de Bruijn-Good graphs, *Discrete Mathematics* 85 (1990) 105-109.

[38] M.L. Liu, Automorphisms of 2-D de Bruijn graphs, *Kexue Tongbao* 2 (1990) 84-87.

[39] M.L. Liu and S.M. Lee, The category of the de Bruijn-Good graphs, *Journal of Graduate School* 1 (1984) 12-15.

[40] J. Plesnik and S. Znam, Strongly geodetic directed graphs, in *Recent Advances in Graph Theory, Proc. Symp.*, Prague, Academia Prague, (1975) 455-465.

[41]S.M. Reddy, J.G. Kuhl, S.H. Hosseini, and H. Lee, On digraphs with minimum diameter and maximum connectivity, *Proc. of 20-th Annual Allerton Conference*, (1982) 1018-1026.

[42] S.M. Reddy, D.K. Pradhan, and J.G. Kuhl, Directed graphs with minimal diameter and maximal connectivity, *School of Engineering Oakland Univ. Tech. Rep.*, 1980.

[43] T. Soneoka, Super edge-connectivity of generalized de Bruijn digraphs, preprint.

[44] T. Soneoka, H. Nakada, and M. Imase, Design of d-connected digraph with a minimum number of edges and a quasiminimal diameter, Preprint.

[45] T. Soneoka, M. Imase, and Y. Manabe, Design of d-connected digraph with a minimum number of edvges and a quasiminimal diameter II, preprint.

[46] Z.X. Wan and M.L. Liu, The automorphisms and homomorphisms of de Bruijn graphs, *Acta Math. Sinica* 22 (1979) 170-177.

[47] D.B. West, *Open problems, SIAM Discrete Math. Newsletter* 1 (3) (1991) 9-12.

4

LINK-CONNECTIVITIES OF
EXTENDED DOUBLE LOOP
NETWORKS

Frank K. Hwang
AT&T Bell Laboratories
Murray Hill, NJ 07974

Wen-Ching Winnie Li*
Department of Mathematics
Pennsylvania State University
University Park, PA 16802

4.1 INTRODUCTION

The notion of *extended double loop networks* (EDLN) was introduced in [5]. Such a network, denoted by $G(n; a, e; b, f)$, is a 2-regular digraph (each node has 2 inlinks and 2 outlinks) with n nodes labelled by the residues $0, 1, \ldots, n-1$, of integers modulo n, and $2n$ links $i \rightarrow ai + e$, $i \rightarrow bi + f$, for $i = 0, 1, \ldots, n-1$. Many 2-regular digraphs popular as topologies for interconnecting networks are special EDLNs. For example, $G(n; 2, 0; 2, 1)$ is the generalized de Bruijn network [6],[9], $G(n; -2, -1; -2, -2)$ is the Imase-Itoh network [7], $G(n; 1, e; 1, f)$ is the usual double loop network [3], and $G(n; 1, 1; 1, f)$ is the FLBH (forward loop backward hop) network [9],[11]. EDLNs are interesting not only because they are a natural generalization of the networks well studied before, but also because their graph structures, despite simple linking patterns, are reasonably complicated due to the "noncommutative" nature of the two types of links (that is, the two paths of length two starting from a node using both types of links do not usually terminate at the same node) and the fact that they are not necessarily Cayley graphs [5]. Our ultimate goal is to investigate various properties of EDLNs to see if certain EDLNs will serve as good interconnection networks. In this paper we study the connectivities of such networks.

---

*This research was supported in part by a grant from NSA no. MDA904-90-H-1021. Part of the work was done when this author was visiting AT&T Bell Laboratories, Murray Hill, NJ.

Ding-Zhu Du and D. Frank Hsu (eds.), Combinatorial Network Theory, 107–124.
© *1996 Kluwer Academic Publishers. Printed in the Netherlands.*

A digraph is said to be *k-link-connected* if there exist k link-disjoint paths from any node to any other node. In this paper we will only be concerned with link connectivity (instead of node-connectivity). Hence we will simply write connectivity for link connectivity. Digraphs considered for interconnection networks are usually 1-connected to allow communication among pairs of nodes, and higher connectivity digraphs provide better fault tolerance in case some links should fail. Clearly, a 2-regular digraph is at most 2-connected since any path from a node must use one of its outlinks. Further, it cannot be 2-connected if it contains a *loop*, that is, a link from a node to itself.

For special 2-regular networks, the following results on connectivity are known. By deriving a finite diameter, Imase-Itoh [6] and Reddy-Pradhan-Kuhl [10] independently showed that $G(n; 2, 0; 2, 1)$ is 1-connected. Since $G(n; 2, 0; 2, 1)$ always contains the loop $0 \rightarrow 0$, it is not 2-connected. Imase-Itoh [7] also showed that $G(n; -2, -2; -2, -1)$ is 1-connected by a similar argument. Du and Hwang [2] proved that it is 2-connected if and only if it has no loops. For $G(n; 1, e; 1, f)$, van Doorn [12] showed that $gcd(e, f, n) = 1$ is necessary and sufficient for it to be 1-connected, and, as observed by Cheng-Hwang-Akyildiz-Hsu [1], in which case the network is 2-connected if and only if it has no loops.

The determination of k-connectivity of an EDLN is a much harder problem. First we recall some facts about EDLNs proved in [5]. The 2-regularity condition imposed on an EDLN $G(n; a, e; b, f)$ is reflected on the parameters as either $gcd(a, n) = gcd(b, n) = 1$ or $gcd(a, n) = gcd(b, n) = 2$ and $e - f$ is odd. Further, up to graph isomorphism, we may always choose parameters of a given EDLN such that $d = gcd(e - f, n)$ divides $a - b$, which we shall so assume. Given an EDLN $G(n; a, e; b, f)$, let

$$n' = \begin{cases} n & \text{if } gcd(a, n) = gcd(b, n) = 1; \\ \text{the largest odd factor of } n & \text{if } gcd(a, n) = gcd(b, n) = 2 \text{ and } e - f \\ & \text{is odd}. \end{cases}$$

Thus we have $gcd(a, n') = gcd(b, n') = 1$ and $d = gcd(e - f, n) = gcd(e - f, n')$. Denote by b^{-1} an integer such that $bb^{-1} \equiv 1 \pmod{n'}$, and by c an integer congruent to ab^{-1} modulo n'. For a node i of $G(n'; a, e; b, f)$, define

$$S(i) = \{i, ci + b^{-1}(e - f), c^2 i + (c + 1)b^{-1}(e - f), \dots, c^{m-1} i + (c^{m-2} + \dots + c + 1)b^{-1}(e - f)\},$$

where m is the smallest positive integer such that

$$c^m i + (c^{m-1} + \dots + c + 1)b^{-1}(e - f) \equiv i \pmod{n'}.$$

Note that in the above ordering of the nodes in $S(i)$, a node j followed by a node k means that the two nodes are related by the congruence equation $aj + e \equiv bk + f \pmod{n'}$. The network $G(n'; a, e; b, f)$ is said to be m-*uniform* if all $S(i)$'s have the same cardinality m.

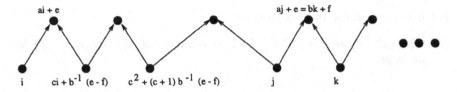

Figure 4.1 nodes in $S(i)$.

In [5] Hwang-Li studied the Hamiltonian property of an EDLN, in particular, they proved the following result on 1-connectivity:

Theorem. *Given an EDLN $G(n; a, e; b, f)$, define n' and d as above. Assume that $G(n'; a, e; b, f)$ is n'/d-uniform. Then $G(n; a, e; b, f)$ is 1-connected if and only if the network $G(d; a, e)$ is 1-connected.*

Here $G(m; q, r)$ denotes the network with m nodes, labelled by the integers modulo m, and m links given by $i \to qi + r$ for each i mod m.

The purpose of this paper is to show

Theorem 4.1.1 *Suppose that $G(n; a, e; b, f)$ is an n'/d-uniform EDLN. Then it is 2-connected if and only if it is 1-connected and has no loops.*

The uniformity condition can be characterized algebraically in terms of the parameters as follows.

Theorem 4.1.2 *Let $G(n; a, e; b, f)$ be an EDLN with $gcd(a, n) = gcd(b, n) = 1$ and $d = gcd(e - f, n)$. Then $G(n; a, e; b, f)$ is n/d-uniform if and only if every prime factor of n/d divides $(a - b)/d$, and 4 divides $a - b$ whenever it divides n/d.*

Observe that the algebraic condition is trivially satisfied when $a = b$. In view of the above theorems, we have immediately the following consequence:

Corollary 4.1.3 *Given an EDLN $G(n; a, e; a, f)$, the following three statements are equivalent:*

(1) it is 2-connected;

(2) it is 1-connected and has no loops;

(3) $G(d; a, e)$, where $d = gcd(e - f, n)$, is 1-connected and $G(n; a, e; a, f)$ has no loops.

The generalized de Bruijn network, the Imase-Itoh network, and the usual double loop network are special EDLNs with $a = b$. Hence our Theorem 4.1.1 is a generalization and unification of the known results on 2-connectivity of special 2-regular networks.

Note that the 1-connectivity of $G(d; a, e)$ and the looplessness of $G(n; a, e; b, f)$ can also be described in terms of simple algebraic conditions on the parameters, as stated in Lemmas 4.2.1 and 4.3.3, respectively. Thus all the conditions concerning the 2-connectivity of an EDLN can be easily checked.

Theorem 4.1.1 is proved in Sections 4.3 and 4.4, here we give the main ideas. The first step is to reduce the proof to the case where $gcd(a, n) = gcd(b, n) = 1$. In this case, let $c = ab^{-1}$ as before. Assuming that the network is not 2-connected, we show that, as a consequence of the uniformity condition, $d = gcd(e - f, n) = 1$ and obtain a set $V' = \{0, 1, c+1, \ldots, c^{k-1} + \ldots + c + 1\}$ of $k+1$, $2 \le k + 1 \le n - 2$, nodes of the network invariant under the map H on $\mathbf{Z}/n\mathbf{Z}$ which sends i to $b^{-1}i + j_0$ with $j_0 \not\equiv 0 \pmod{n}$. In case $a = b$, that is, $c = 1$, the above is possible only when $b \equiv -1 \pmod{n}$. This cannot happen because the loopless condition together with $d = 1$ implies $b \not\equiv -1 \pmod{n}$. The case of general c is a lot more complicated. To create an analogous situation, consider $C = gcd(c - 1, b - 1, n)$ so that $c \equiv b \equiv 1 \pmod{C}$. We show that $C \ge 3$ and C divides $k + 1$. Then we partition V' into C subsets of the same cardinality $s + 1 = (k + 1)/C$ according to the residues modulo C. The map H permutes these subsets. By studying the action of H in details, we prove the existence of another set $Y = \{c^{C(t-1)} + c^{C(t-2)} + \ldots + c^C + 1$

$\pmod{n/C} : 0 \le t \le s\}$ of size $s + 1$ which is invariant under the map on $\mathbf{Z}/(n/C)\mathbf{Z}$ sending i to $c^C i + 1$. The n-uniformity of the network imposes a condition on c as described in Theorem 4.1.2, which in turn implies that Y must have cardinality n/C, which is not the case.

4.2 PROOF OF THEOREM 4.1.2

We shall need the following result from [4] (first proved in [8] under a different context) characterizing the connectivity of $G(m; q, r)$.

Lemma 4.2.1 $G(m; q, r)$ is 1-connected if and only if $\gcd(r, m) = 1$, all prime factors of m divides $q - 1$, and 4 divides $q - 1$ whenever 4 divides m.

Now we prove Theorem 4.1.2. By definition, $G(n; a, e; b, f)$ is n/d-uniform if and only if for each node i, the set

$$S(i) = \{i, \ ci + b^{-1}(e - f), \ c^2 i + (c + 1)b^{-1}(e - f), \ldots,$$

$$c^j i + (c^{j-1} + c^{j-2} + \ldots + c + 1)b^{-1}(e - f), \cdots\}$$

has cardinality n/d. In other words, for every integer i, n/d is the smallest positive integer m such that

$$c^m i + (c^{m-1} + c^{m-2} + \ldots + c + 1)b^{-1}(e - f) \equiv i \pmod{n},$$

i.e.,

$$(c^{m-1} + c^{m-2} + \ldots + c + 1)\left[(c - 1)i + b^{-1}(e - f)\right] \equiv 0 \pmod{n}.$$

Since d divides $c - 1$ and $e - f$, the above congruence equation is equivalent to

$$(c^{m-1} + c^{m-2} + \ldots + c + 1)\left[\frac{c - 1}{d}i + b^{-1}\frac{e - f}{d}\right] \equiv 0 \pmod{\frac{n}{d}}.$$

First we prove sufficiency. As every prime factor of $\frac{n}{d}$ divides $\frac{a-b}{d}$, hence $\frac{c-1}{d}$ (since b is prime to n), and $b^{-1}\frac{e-f}{d}$ is prime to $\frac{n}{d}$, $G(n; a, e; b, f)$ will be n/d-uniform if we can show that n/d is the smallest positive integer m such that

$$c^{m-1} + c^{m-2} + \ldots + c + 1 \equiv 0 \pmod{\frac{n}{d}}.$$

Thus the sufficiency part of the theorem follows from

Lemma 4.2.2 Let k be a positive integer and let c be a positive integer prime to k. Suppose that every prime factor of k divides $c - 1$ and 4 divides $c - 1$ whenever 4 divides k. Then the smallest positive integer m such that

$$c^{m-1} + c^{m-2} + \ldots + c + 1 \equiv 0 \pmod{k}$$

is k.

Proof. The lemma is trivial when $k = 1$. Assume $k > 1$. Let p be a prime factor of k. Suppose $c^{m-1} + \ldots + c + 1 \equiv 0 \pmod{k}$. Call the highest power of p dividing x the *order* of p in x. Denote by α and β the order of p in $c - 1$ and m, respectively. Then $\alpha \geq 1$ by hypothesis. We study the order of p in

$$c^m - 1 = (c - 1)(c^{m-1} + \ldots + c + 1) .$$

It can be shown inductively on β that the order of p in $c^m - 1$ is $\alpha + \beta$ unless $p = 2$ and $\alpha = 1$. Since k divides $c^{m-1} + \ldots + c + 1$ by assumption, β is at least the order of p in k. Write $k = 2^{\gamma} k'$, where $\gamma \geq 0$ and k' is odd. We have shown that k' divides m and 2^{γ} divides m if $\gamma \geq 2$ because in which case we have $\alpha \geq 2$ by hypothesis. If $\gamma = 1$, then k is even and c is odd, hence m must be even in order that $c^{m-1} + \ldots + c + 1$ be even. Thus m is a multiple of k in all cases, and as analyzed above, the smallest such m is k. $\qquad\square$

Next we prove necessity. Assume that $G(n; a, e; b, f)$ is n/d-uniform. Then

$$S(0) = \{0, b^{-1}(e - f), (c + 1)b^{-1}(e - f), \ldots, (c^{n/d-2} + \ldots + c + 1)b^{-1}(e - f)\}$$

has cardinality n/d. Note that every node in $S(0)$ is a multiple of d, hence $S(0)$ consists of the residues mod n which are multiples of d. By listing the nodes in $S(0)$ as ordered and adjoining a link from the ith node to the $i + 1^{\text{st}}$ for $i = 1, \ldots, n/d$ and a link from the last node to the first, we obtain a digraph which, after removing the common multiple d from all nodes, is nothing but $G(n/d; c, b^{-1}(e-f)/d)$. Hence $|S(0)| = n/d$ implies that $G(n/d; c, b^{-1}(e-f)/d)$ is 1-connected. By Lemma 4.2.1, we have

(A) all prime factors of n/d divide $c - 1$, and 4 divides $c - 1$ whenever it divides n/d.

As $c \equiv ab^{-1} \pmod{n}$ and b is prime to n, we may replace $c - 1$ by $a - b$ in (A). It remains to show that the first statement in (A) holds with $c - 1$ replaced by $(a-b)/d$, that is, by $(c-1)/d$. Suppose that there is a prime factor p of n/d which does not divide $(c - 1)/d$. Then there is an integer i such that $i(c-1)/d + b^{-1}(e-f)/d \equiv 0 \pmod{p}$, hence $D = \gcd((c-1)i + b^{-1}(e-f), n)$ is a multiple of pd. Consider the set $S(i)$. As explained above, its cardinality is the smallest integer m such that

$$(c^{m-1} + \ldots + c + 1)((c - 1)i + b^{-1}(e - f)) \equiv 0 \pmod{n} ,$$

i.e.,

$$c^{m-1} + \ldots + c + 1 \equiv 0 \pmod{n/D} .$$

Condition (A) and Lemma 4.2.2 imply that $m = n/D$. Thus $|S(i)| = n/D \leq n/dp < n/d$, contradicting the n/d-uniformity assumption. This proves Theorem 4.1.2.

Contained in the above proof is the following

Corollary 4.2.3 *Let* $G(n; a, e; b, f)$ *be an EDLN with* $gcd(a, n) = gcd(b, n) = 1$ *and* $d = gcd(e - f, n)$. *If* $G(n; a, e; b, f)$ *is* n/d-uniform, *then* $G(n/d; c, b^{-1}(e - f)/d)$ *is 1-connected.*

4.3 PREPARATION FOR THE PROOF OF THEOREM 4.1.1

Clearly, only the sufficiency part of Theorem 4.1.1 requires a proof. Let $G(n; a, e; b, f)$ be an EDLN. As explained in Introduction, the 2-regularity condition of the network is described by two algebraic conditions, which lead to the definition of n'. Our first task is to reduce the proof to the case where $gcd(a, n) = gcd(b, n) = 1$, that is, $n' = n$.

The line graph G' of a digraph G is a digraph whose nodes are the links of G and there is a link from one node going to another node of G' if and only if the two corresponding links in G are consecutive. The following two statements were proved in [5]:

Lemma 4.3.1 *Let* $G(n; a, e; b, f)$ *be an EDLN with* $n' < n$. *Then* $G(n; a, e; b, f)$ *is the line graph of the EDLN* $G(n/2; a, e; b, f)$.

Lemma 4.3.2 *Let* $G(n; a, e; b, f)$ *be as above. Then* $G(n; a, e; b, f)$ *is 1-connected if and only if* $G(n'; a, e; b, f)$ *is.*

The loopless condition can be easily described as follows.

Lemma 4.3.3 *An EDLN* $G(n; a, e; b, f)$ *has no loops if and only if* $gcd(a-1, n)$ *does not divide* e *and* $gcd(b - 1, n)$ *does not divide* f.

Proof. A link $i \to ai + e$ is a loop if and only if $i \equiv ai + e$ (mod n), which is possible if and only if $gcd(a - 1, n)$ divides e. The same argument holds for a loop of type $i \to bi + f$. $\qquad\qquad\qquad\qquad\qquad\qquad\qquad\qquad\qquad\qquad\qquad\qquad\quad$ $\square$

We can now prove the desired reduction.

Lemma 4.3.4 *If Theorem 4.1.1 holds for EDLNs with $n' = n$, then it also holds for those with $n' < n$.*

Proof. Let $G(n; a, e; b, f)$ be an EDLN with $n' < n$. The 1-connectivity of $G(n'; a, e; b, f)$ follows from that of $G(n; a, e; b, f)$ as implied by Lemma 4.3.2. Further, since a and b are even, we have $gcd(a - 1, n) = gcd(a - 1, n')$ and $gcd(b - 1, n) = gcd(b - 1, n')$. We see from Lemma 4.3.3 that $G(n'; a, e; b, f)$ has no loops since $G(n; a, e; b, f)$ has none. Thus by assumption, the network $G(n'; a, e; b, f)$ is 2-connected. $\qquad\qquad\qquad\qquad\qquad\qquad\qquad\qquad\qquad\quad$ $\square$

Write $n = 2^t n'$. By Lemma 4.3.1, $G(n; a, e; b, f)$ is the t-step line graph of $G(n'; a, e; b, f)$. As the line graph of a 2-connected digraph has two node-disjoint paths, consequently, it has two link-disjoint paths, in other words, the line graph is also 2-connected. This proves that $G(n; a, e; b, f)$ is 2-connected.

We end this section by proving a theorem describing the graph structure of a 1-connected 2-regular digraph with no loops which is not 2-connected.

Theorem 4.3.5 *Let G be a 2-regular 1-connected digraph with no loops. Suppose that the graph X obtained from G by removing an edge $v \to u$ is not 1-connected. Then there are two disjoint subgraphs U, V of X, containing u, v, respectively, such that X is the union of U and V together with a link $x \to y$ from U to V.*

Proof. As G is 1-connected, every node in X can reach v via a path in X. Denote by U the subgraph of X induced by the set of nodes in X which can reach u via paths in X. Then $v \notin U$ because X is not 1-connected. $\qquad\quad$ $\square$

Since G is 2-regular and contains no loop, U contains at least 2 nodes. Each node in U has two inlinks except u which has only one inlink, thus there are $2|U| - 1$ inlinks at the nodes in U. Since the starting nodes of these inlinks are in U, these inlinks are also the outlinks from nodes in U. Therefore all nodes in U have 2 outlinks except one, say, x, which has only one outlink in U. Let

y be the endpoint of the outlink from x which is not in U. Denote by V the subgraph of X induced by nodes in X reachable from y. Since y reaches v, V contains the node v. Every node in V has 2 outlinks except v which has only one outlink. So there are $2|V| - 1$ outlinks in V. As the endpoints of these outlinks are in V, there are also $2|V| - 1$ inlinks in V. As the node y has one inlink $x \to y$ not in V, we know that all nodes in V except y have 2 inlinks in V, and the node y has only 1 inlink in V. Again, 2-regularity together with looplessness of G implies that V has at least 2 nodes. If there is a node w of X outside $V \cup U$, let p be a path in X from w to v. Since w is outside U, no node on p lies in U (otherwise w can reach u and thus lies in U). Starting from v, trace backwards along the path p. Let z be the last node on p such that the path from z to v are in V. Then $z \neq w$ since w is outside V. Let z' be the predecessor of z on p. Then z' is outside $V \cup U$. This shows that the node z in V has an inlink $z' \to z$ from a node not in $V \cup U$, a contradiction. Therefore the nodes of X are partitioned into 2 parts, those in U and those in V; the union of the graph U and graph V together with the link $x \to y$ is the graph X.

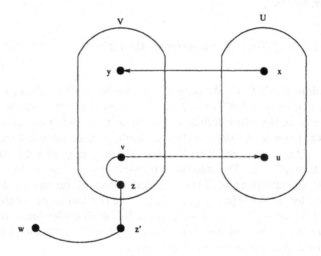

Figure 4.2   a node outside of $V \cup U$.

4.4   PROOF OF THEOREM 4.1.1

Only the sufficiency part needs to be proved. Let $G = G(n; a, e; b, f)$ be an EDLN which is 1-connected and loopless. In view of Lemma 4.3.4, we may assume that $gcd(a, n) = gcd(b, n) = 1$ and hence $n' = n$. Let $d = gcd(e - f, n)$. Suppose that G is n/d-uniform. Thus for each node i of G, the set $S(i)$ defined in Introduction has cardinality n/d, consisting of the nodes in G which are congruent to $i \bmod d$ since $c \equiv ab^{-1} \pmod{n}$ and d divides $a - b$ by the choice of parameters. Further, the outneighbors of the nodes in $S(i)$ are the nodes in $S(ai + e) = S(bi + f)$. We want to show that G is 2-connected. Our method is to suppose otherwise and derive a contradiction. Keep the same notation as in Theorem 3. Denote by y' the node in V which has an outlink to y. Without loss of generality, we may assume

$$y \equiv by' + f \pmod{n} .$$

We begin by showing

Theorem 4.4.1 *If G is not 2-connected, then $d = 1$.*

Proof. Consider $S(y') \cap V$. Arrange the nodes in $S(y')$ in the order y', $cy' + b^{-1}(e - f)$, $c^2 y' + (c + 1)b^{-1}(e - f), \ldots$. Then the last node in this list is x since $x \to y$ is the other inlink to y. As $x \in U$, we have $|S(y') \cap V| < n/d$. Further, if we let y'' denote the last node in the above list such that y', $cy' + b^{-1}(e - f), \ldots, y''$ are all in V, then there is only one outlink from y'' lying in V, thus $y'' = v$. The missing outlink is $v \to u$. Let u' be the node in U which has an outlink to u. Then the node after y'' on the list above is u', then followed by $cu' + b^{-1}(e - f), \ldots$. Let z be the last node on this list such that $u', cu' + b^{-1}(e - f), \ldots, z$ are all in U. Since all nodes but u in U have 2 inlinks, there is only one outlink from z which is in U. This shows that $z = x$. We have shown that the nodes in $S(y')$ listed as

$$y', cy' + b^{-1}(e - f),\ c^2 y' + (c + 1)b^{-1}(e - f), \ldots, v, u'$$

$$= \ cv + b^{-1}(e - f),\ cu' + b^{-1}(e - f), \ldots, x$$

split into two parts, from y' to v lying in V, and from u' to x lying in U. Further, arrange the nodes in $S(y)$ in the order

$$y = by' + f,\quad b(cy' + b^{-1}(e - f)) + f, \ldots, bv + f,\quad bu' + f, \ldots, bx + f , \quad (4.1)$$

then they also split into two parts, from y to $bv + f$ in V and from $bu' + f$ to $bx + f$ in U. In particular,

$$|S(y') \cap V| = |S(y) \cap V| < n/d .$$

□

If $d > 1$, the 1-connectedness of G implies that of $G(d; a, e)$, in particular, $ay' + e \not\equiv y' \pmod{d}$. Hence $S(y) = S(ay' + e) \neq S(y')$. Then every node in $S(y) \cap V$ has 2 outlinks in V. Further, since $S(ay + e) \neq S(y)$, every node in $S(ay + e) \cap V$ has 2 inlinks in V. This then forces $|S(y) \cap V| = |S(y)| = n/d$, a contradiction. Therefore $d = gcd(e - f, n) = 1$.

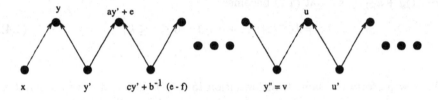

Figure 4.3   nodes in $S(y')$.

Lemma 4.4.2 *If G is not 2-connected, then $gcd(c^{i-1} + \ldots + c + 1, n) = gcd(i, n)$ for $i \geq 1$.*

Proof. We know from Theorem 4.4.1 that $d = 1$. Thus G is n-uniform, hence every prime factor of n divides $c - 1$ and 4 divides $c - 1$ whenever it divides n by Theorem 4.1.2. As in the proof of Lemma 4.3.3 of [5], this condition in turn implies that for each prime factor p of n, the highest power of p dividing $c^{i-1} + \ldots + c + 1$ is exactly the highest power dividing i. This proves the lemma. □

When the elements of $S(y)$ as listed in (4.1) are expressed in terms of y, they are

$$y, \quad cy + e - cf, \quad c^2y + (c+1)(e - cf), \ldots, c^my + (c^{m-1} + \ldots + c + 1)(e - cf), \ldots .$$

Suppose that V has $k + 1$ elements, where $k \geq 1$. We have by Theorem 4.3.5,

$$V = S(y) \cap V = \{c^iy + (c^{i-1} + \ldots + c + 1)(e - cf) : 0 \leq i \leq k\} \quad (4.2)$$

$$= \ S(y') \cap V = \{b^{-1}(c^i y + (c^{i-1} + \ldots + c + 1)(e - cf)) - b^{-1}f : \\ 0 \le i \le k\} \ . \tag{4.3}$$

Here we used the convention that $c^{i-1} + \ldots + c + 1 = 0$ if $i = 0$. To simplify our notations, write

$$c^i y + (c^{i-1} + \ldots + c + 1)(e - cf) = (c^{i-1} + \ldots + c + 1)((c - 1)y + e - cf) + y \ .$$

Noting that

$$gcd((c - 1)y + e - cf, n) = gcd((c - 1)(y - f) + e - f, n) = 1$$

since every prime factor of n divides $c - 1$ and $gcd(e - f, n) = 1$, we transform the elements in V by first subtracting y and then multiplying by the inverse of $(c - 1)y + e - cf$ so that (4.2) becomes

$$V' = \{c^{i-1} + \ldots + c + 1 : \ 0 \le i \le k\} \ . \tag{4.4}$$

We now perform the same transformation for (4.2). Let j_0 denote the number mod n obtained by transforming $y' = b^{-1}y - b^{-1}f$ in (4.2), that is,

$$j_0 \equiv (b^{-1}y - b^{-1}f - y)((c - 1)y + e - cf)^{-1} \quad (\text{mod } n) \ .$$

Then the number mod n obtained by transforming any other node

$$b^{-1}(c^i y + (c^{i-1} + \ldots + c + 1)(e - cf)) - b^{-1}f$$

$$= \ b^{-1}(c^{i-1} + \ldots + c + 1)((c - 1)y + e - cf) + b^{-1}y - b^{-1}f$$

is $b^{-1}(c^{i-1} + \ldots + c + 1) + j_0 \pmod{n}$. Therefore from the two expressions (4.2) and (4.3) of the same set V we obtain

$$V' = \{c^{i-1} + \ldots + c + 1 : \ 0 \le i \le k\} = \{b^{-1}(c^{i-1} + \ldots + c + 1) + j_0 : \ 0 \le i \le k\} \ . \tag{4.5}$$

Since $y \equiv b^{-1}(y - f) \pmod{n}$ implies a loop at the node y, we have $j_0 \not\equiv 0$ $\pmod{n}$. Further, from

$$((c - 1)y + e - cf)j_0 \equiv -b^{-1}((b - 1)y + f) \pmod{n}$$

and the fact that b and n are relatively prime, $gcd(b - 1, n)$ cannot divide j_0, or else it would divide f, contradicting Lemma 4.3.3.

Denote by H the map on $\mathbf{Z}/n\mathbf{Z}$ sending x to $b^{-1}x + j_0$. Then (4.4) says that V' is invariant under the map H.

The following lemma will be used repeatedly:

Lemma 4.4.3 *Let m be a divisor of n and $T = \{c^{i-1}+\ldots+c+1 \pmod{m} : 0 \le i \le l\}$. If T is invariant under the map $h(x) = x + j$ on $\mathbf{Z}/m\mathbf{Z}$ with $j \not\equiv 0 \pmod{m}$, then $T = \mathbf{Z}/m\mathbf{Z}$.*

Proof. Let $J = gcd(j, m)$. By assumption, $J < m$ and T contains the additive subgroup A generated by j, which has m/J elements. As $gcd(c^{i-1} + \ldots + c + 1, m) = gcd(i, m)$, the group A contains $c^{J-1} + \ldots + c + 1$. Therefore T contains $0, 1, c+1, \ldots, c^{J-2} + \ldots + c + 1$. Further, these elements represent distinct cosets of A since the gcd of m and the difference of two distinct elements is less than J. Therefore T contains the J distinct cosets of A in $\mathbf{Z}/m\mathbf{Z}$, that is, T is equal to $\mathbf{Z}/m\mathbf{Z}$. $\square$

Let $B = gcd(b-1, n)$. As remarked before, B does not divide j_0. In particular, $B > 1$.

Lemma 4.4.4 *B divides the cardinality of V', which is $|V| = k+1$.*

Proof. Let q be the quotient of $k+1$ divided by B and let r be the remainder. Note that $gcd(c^{i-1} + \ldots + c + 1, B) = gcd(i, B)$ by Lemma 4.4.2. If $r > 0$, then the set $W_i = \{z \in V' : z \equiv c^{i-1} + \ldots + c + 1 \pmod{B}\}$ has cardinality $q+1$ for $i = 0, \ldots, r-1$, and cardinality q for $i = r, \ldots, B-1$. Since the map $H(x) = b^{-1}x + j_0$ sends V' to V' and $b \equiv 1 \pmod{B}$ H permutes the sets $W_0, \ldots, W_{r-1}$. In particular, H induces a map $\bar{H}(x) = x + j_0 \pmod{B}$ on $\mathbf{Z}/B\mathbf{Z}$ which leaves the set $T = \{c^{i-1} + \ldots + c + 1 \pmod{B} : 0 \le i \le r-1\}$ invariant. As $j_0 \not\equiv 0 \pmod{B}$ we have $T = \mathbf{Z}/B\mathbf{Z}$ by Lemma 4.4.3, which is impossible. $\square$

Next let $C = gcd(c-1, B)$. Since $B > 1$, B divides n and every prime divisor of n divides $c - 1$, we have $C > 1$. Write $|V'| = k + 1 = C(s+1)$ with $s \ge 0$. Note that $c \equiv 1 \pmod{C}$. Partition the elements in V' into C sets according to the residues modulo C. More precisely, for $0 \le i \le C - 1$, let

$$V_i' = \{c^{i-1} + \ldots + c + 1, \quad c^{C+i-1} + \ldots + c + 1, \ldots, c^{sC+i-1} + \ldots + c + 1\},$$

Then V' is a disjoint union of $V'_0, V'_1, \ldots, V'_{C-1}$. As $gcd(c^{C-1} + \ldots + c + 1, n) = C$ by Lemma 4.4.2, write $c^{C-1} + \ldots + c + 1 = C\alpha$, then α is prime to n. The set V'_i can also be expressed as

$$V'_i = \{c^{i-1} + \ldots + c + 1 + (c^{C(t-1)} + c^{C(t-2)} + \ldots + c^C + 1)c^i C\alpha : \ 0 \le t \le s\},$$

The map $H: \ x \to b^{-1}x + j_0$ sends V'_i to $V'_{i'}$, where $i' \equiv b^{-1}i + j_0 \equiv i + j_0$ (mod C). Denote by i_0 the residue of j_0 modulo C. Then $i' \equiv i + i_0$ (mod C). More precisely, for $0 \le i \le C - 1 - i_0$ we have $i' = i + i_0$, and for $C - i_0 \le i \le C - 1$ we have $i' = i + i_0 - C$. So, for $0 \le i \le C - 1$,

$$
\begin{aligned}
H(c^{i-1} + \ldots + c + 1) &= b^{-1}(c^{i-1} + \cdots + c + 1) + j_0 \\
&= c^{i'-1} + \ldots + c + 1 + J_{i'}c^{i'}C\alpha \in V'_{i'} ,
\end{aligned}
$$

and for $0 \le t \le s$,

$$
\begin{aligned}
&H(c^{i-1} + \ldots + c + 1 + c^i C\alpha(c^{C(t-1)} + \ldots + c^C + 1)) \\
&= b^{-1}c^i C\alpha(c^{C(t-1)} + \ldots + c^C + 1) + c^{i'-1} + \ldots + c + 1 + J_{i'}c^{i'}C\alpha \in V'_{i'} .
\end{aligned}
$$

Therefore, for each $0 \le t \le s$ there is a unique $0 \le t' \le s$ such that

$$
\begin{aligned}
&b^{-1}c^i C\alpha(c^{C(t-1)} + \ldots + c^C + 1) + J_{i'}c^{i'}C\alpha \\
&\equiv (c^{C(t'-1)} + \ldots + c^C + 1)c^{i'}C\alpha \pmod{n},
\end{aligned}
$$

or equivalently,

$$b^{-1}c^{i-i'}(c^{C(t-1)} + \ldots + c^C + 1) + J_{i'} \equiv c^{C(t'-1)} + \ldots + c^C + 1 \pmod{n/C} .$$

Let

$$Y = \{c^{C(t-1)} + \ldots + c^C + 1 \pmod{n/C} : \ 0 \le t \le s\} .$$

We have shown that Y is invariant under the map H_i on $\mathbf{Z}/\frac{n}{C}\mathbf{Z}$ given by

$$H_i(x) = b^{-1}c^{i-i'}x + J_{i'} .$$

For $0 \le i_1, i_2 \le C - 1 - i_0$, we have $i_1 - i'_1 = -i_0 = i_2 - i'_2$ and Y is invariant under the map $H_{i_2}^{-1}H_{i_1}$, which is given by

$$H_{i_2}^{-1}H_{i_1}(x) = x + bc^{i_0}(J_{i'_1} - J_{i'_2}) .$$

As $|Y| = s + 1 = |V'|/C < n/C$, we conclude from Lemma 4.4.3 that the $J_{i'}$ are the same for $0 \le i \le C - 1 - i_0$, denote this common value by J. Similarly, the $J_{i'}$ are the same for $C - i_0 \le i \le C - 1$, denote this common value by J'.

Next we rule out the case $i_0 = 0$. Suppose otherwise, that is, $i_0 = 0$. Then each V_i' is invariant under H, and by our analysis above, $H(0) = j_0 = JC\alpha = c^{t_0 C - 1} + \ldots + c + 1$ for some $1 \le t_0 \le s$ since $j_0 \not\equiv 0 \pmod{n}$, and $H(1) = b^{-1} + j_0 = 1 + JcC\alpha$, which yields

$$b^{-1} \equiv 1 + (c - 1)JC\alpha \pmod{n} .$$

With this information, we get, for $0 \le i \le k - 1$,

$$
\begin{aligned}
H(c^i + \ldots + c + 1) &\equiv b^{-1}(c^i + \cdots + c + 1) + j_0 \\
&\equiv c^i + \ldots + c + 1 + (c^i + 1 - 1)JC\alpha + JC\alpha \\
&\equiv c^i + \ldots + c + 1 + c^{i+1}(c^{t_0 C - 1} + \ldots + c + 1) \\
&\equiv c^{t_0 C + i} + \ldots + c + 1 \pmod{n} .
\end{aligned}
$$

In particular, when $i = (s + 1 - t_0)C - 1$, we have $c^{(s+1-t_0)C-1} + \ldots + c + 1 \in V'$. But its image under H is $c^{(s+1)C-1} + \ldots + c + 1 = c^k + \cdots + c + 1$, which is not in V', a contradiction.

We summarize the above discussion in

Lemma 4.4.5 *The set Y is invariant under the following two maps on $\mathbf{Z}/\frac{n}{C}\mathbf{Z}$:*

$$h_1(x) = b^{-1}c^{-i_0}x + J \quad and \quad h_2(x) = b^{-1}c^{C-i_0}x + J' ,$$

where $C > i_0 > 0$, and $J, J' \in \mathbf{Z}/\frac{n}{C}\mathbf{Z}$ are such that

$$
\begin{aligned}
H(c^{i-1} + \ldots + c + 1) &\equiv c^{i+i_0-1} + \ldots + c + 1 + Jc^{i+i_0}C\alpha \\
&\qquad\qquad for \ \ 0 \le i \le C - 1 - i_0 , \\
&\equiv c^{i+i_0-C-1} + \ldots + c + 1 + J'c^{i+i_0-C}C\alpha \\
&\qquad\qquad for \ \ C - i_0 \le i \le C - 1 . \quad (4.6)
\end{aligned}
$$

To continue, we observe that $C > 2$. Indeed, since $C = \gcd(c - 1, B) = \gcd(c - 1, b - 1, n)$ and, by assumption, all odd prime factors of n divide $c - 1$ and 4 divides $c - 1$ if 4 divides n, $C = 2$ would imply $\gcd(b - 1, n) = 2$. This in turn implies that n is even and f is odd since $\gcd(b - 1, n)$ does not divide f (as there is no loop). On the other hand, 2 divides $c - 1 = ab^{-1} - 1$, thus a is also odd and consequently e is odd. But then $\gcd(e - f, n) \ge 2$, a contradiction.

Therefore $C \geq 3$. Next we compare J and J'. It follows from (4.6) that

$$H(0) \equiv j_0 \equiv c^{i_0-1} + \ldots + c + 1 + Jc^{i_0}C\alpha ,$$

$$\begin{aligned} H(c^{C-2-i_0} + \ldots + c + 1) &\equiv b^{-1}(c^{C-2-i_0} + \ldots + c + 1) + j_0 \\ &\equiv c^C - 2 + \ldots + c + 1 + c^{C-1}JC\alpha, \end{aligned}$$

$$H(c^{C-1-i_0} + \ldots + c + 1) \equiv b^{-1}(c^{C-1-i_0} + \ldots + c + 1) + j_0 \equiv 0 + J'C\alpha ,$$

which implies

$$b^{-1}c^{C-1-i_0} \equiv -(c^{C-2} + \ldots + c + 1) - c^{C-1}JC\alpha + J'C\alpha \quad (\text{mod } n) . \quad (4.7)$$

On the other hand, since $C \geq 3$, we have either $1 \leq C-1-i_0$ or $C-i_0 < C-1$, thus we get either

$$H(1) \equiv b^{-1} + j_0 \equiv c^{i_0} + \ldots + c + 1 + c^{i_0+1}JC\alpha$$

or

$$H(c^{C-i_0} + \ldots + c + 1) \equiv b^{-1}(c^{C-i_0} + \cdots^c + 1) + j_0 \equiv 1 + J'cC\alpha,$$

which yields

$$b^{-1} \equiv c^{i_0} + (c-1)c^{i_0}JC\alpha, \quad \text{or} \quad b^{-1} \equiv c^{i_0-C} + (c-1)c^{i_0-C}J'C\alpha \quad (\text{mod } n).$$

When combined with (4.7), we obtain

$$J' \equiv 1 + c^C J \quad (\text{mod } n/C)$$

in both cases. Therefore, the map h_2 can be expressed as

$$h_2(x) = b^{-1}c^{C-i_0}x + 1 + c^C J = c^C h_1(x) + 1 .$$

We record the above discussion in

Theorem 4.4.6 *The set* $Y = \{c^{C(t-1)} + \ldots + c^C + 1 \pmod{n/C} : 0 \leq t \leq s\}$ *is invariant under the map* $h(= h_2 \circ h_1^{-1})$ *on* $\mathbf{Z}/\frac{n}{C}\mathbf{Z}$ *given by*

$$h(x) = c^C x + 1 .$$

Finally, we examine the action of h on Y. It sends $c^{C(t-1)} + \ldots + c^C + 1$ to $c^{Ct} + \ldots + c^C + 1$ for $0 \leq t \leq s-1$, and it sends $c^{C(s-1)} + \ldots + c^C + 1$ to $c^{Cs} + \ldots + c^C + 1$. In order that $h(Y) = Y$, we must have

$$c^{Cs} + \ldots + c^C + 1 \equiv 0 \pmod{n/C},$$

or equivalently,

$$0 \equiv (c^{Cs} + \ldots + c^C + 1)(c^{C-1} + \cdots + c + 1) \equiv c^C(s+1) - 1 + \ldots + c + 1 \pmod{n}.$$

But $gcd(c^{C(s+1)-1} + \ldots + c + 1, n) = gcd(C(s+1), n) = gcd(|V|, n)$, which is less than n, a contradiction. This completes the proof of Theorem 4.1.1.

4.5 CONCLUSION

The class of extended double loop networks includes the much studied 2-regular networks in the literature, such as generalized de Bruijn networks, Imase-Itoh networks, and double loop networks. The determination of k-connectivity for an EDLN $G = G(n; a, e; b, f)$ is a difficult problem, even for 1-connectivity. In [5] a simple criterion for 1-connectivity of G was given under a uniformity assumption explained in Introduction. In this paper we show that under the same assumption, the network is 2-connected if and only if it is 1-connected and loopless. It is still unknown if the statement holds unconditionally.

The uniformity condition can also be described algebraically in terms of the parameters of the network G, as shown in Theorem 4.1.2. In particular, it is trivially satisfied when $a = b$, which is the case for the well-studied networks mentioned above. Therefore our result is a generalization and unification of the previously known results on the connectivity of special 2-regular networks.

REFERENCES

[1] Y. Cheng, F. K. Hwang, I. F. Akyildiz and D. F. Hsu, "Routing Algorithms for Double Loop Networks," *Inter. J. Found. Comput. Sci.*

[2] D. Z. Du and F. K. Hwang, "Generalized de Bruijn Digraphs," *Networks* 18, 27-38, 1988.

[3] M. A. Fiol, M. Valero, J. L. A. Yebra, I. Alegre and T. Lang, "Optimization of Double-Loop Structures for Local Networks," *Proc. XIX Int. Symp. MIMI '82*, Paris, 1982, pp. 37-41.

[4] F. K. Hwang, "The Hamiltonian Property of Linear Functions," *Oper. Res. Letters* **6**, 125-127, July 1987.

[5] F. K. Hwang and W.-C. W. Li, "Hamiltonian Circuits for 2-Regular Interconnection Networks," in *Networks Optimization*, Ed: D. Z. Du and P. Pardalos, World Scientific, River Edge, NJ.

[6] M. Imase and M. Itoh, "Design to Minimize Diameter on Building-Block Network," *IEEE Trans. Comput.* **C-30**, 439-442, June 1981.

[7] M. Imase and M. Itoh, "A Design for Directed Graphs with Minimum Diameters," *IEEE Trans. Comput.* **C-32**, 782-784, August 1983.

[8] D. E. Knuth, *The Art of Computer Programming*, Vol. 2, Addison-Wesley, Reading, MA 1972.

[9] C. S. Raghavendra, M. Gerla and A. Avienis, "Reliable Loop Topologies for Large Local Computer Networks," *IEEE Trans. Comput.* **C-34**, 46-54, January 1985.

[10] S. M. Reddy, D. K. Pradhan and J. G. Kuhl, "Direct Graphs with Minimum Diameter and Maximal Connectivity," School of Eng., Oakland Univ. Tech. Rep., July 1980.

[11] C. K. Wong and D. Coppersmith, "A Combinatorial Problem Related to Multinodule Memory Organizations," *J. Assoc. Comput. Mach.* **21**, 392-402, July, 1974.

[12] E. A. van Doorn, "Connectivity of Circulant Digraphs," *J. Graph Theory* **10**, 9-14, 1986.

5

DISSEMINATION OF INFORMATION IN INTERCONNECTION NETWORKS (BROADCASTING & GOSSIPING)

Juraj Hromkovič
Institut für Informatik und Praktische Mathematik
Universität zu Kiel, 24098 Kiel, Germany

Ralf Klasing
Burkhard Monien
Regine Peine
Department of Mathematics and Computer Science
University of Paderborn, 33095 Paderborn, Germany

5.1 INTRODUCTION

Considerable attention in recent theoretical computer science is devoted to parallel computing. Here, we would like to present a special part of this large topic, namely, the part devoted to an abstract study of the dissemination of information in interconnection networks. The importance of this research area lies in the fact that the ability of a network to effectively disseminate information is an important qualitative measure for the suitabilty of the network for parallel computing. This follows simply from the observation that the communication among processes working in parallel is one of the main parts of the whole parallel computation. So, the effectivity of information exchange among processors essentially influences the effectivity of the whole computation process.

The main aims of this work are the following:

1. To provide an easily readable introduction (suitable also for undergraduate students) to the research area dealing with the dissemination of information in distinct interconnection networks.

125

Ding-Zhu Du and D. Frank Hsu (eds.), Combinatorial Network Theory, 125–212.
© *1996 Kluwer Academic Publishers. Printed in the Netherlands.*

2. To explain some of the basic proof techniques and ideas leading to some important results of this area.

3. To give a survey of established results, especially for broadcasting and gossiping in the extensively studied telegraph and telephone communication mode, and to formulate some open problems interesting for this research area.

The structure of this article follows the aims stated above. The first section introduces this research area. The basic definitions are given and the fundamental, simple observations concerning the relations among the complexity measures defined are carefully explained. This section is devoted to people who have never worked in this area and can be skipped by anybody who is familiar with this topic. The notation fixed here is the usual one used in the literature.

The second section is devoted to broadcasting, and it presents some of the main techniques and results connected to broadcast problems in the one–way (telegraph) communication mode.

The third section is devoted to gossiping in the one–way (telegraph) communication mode and in the two–way (telephone) communication mode. It provides also some basic ideas, a survey of the known results, and the formulation of open problems.

The last section provides a short survey of broadcasting and gossiping in other communication modes. It also discusses other possibilities than the number of communication rounds to measure the complexity of information dissemination.

Finally, we give a list of all publications devoted to this topic which are currently known to us.

5.1.1   Motivation and Definitions

A lot of work has been done in recent years in the study of the properties of interconnection networks in order to find the best communication structures for parallel and distributed computing. An important feature characterizing the "quality" (suitability) of an interconnection network for parallel computing is the ability to effectively disseminate the information among its processors. Thus, the problem of dissemination of information has been investigated for most of the interconnection networks considered in parallel computing.

There are three main problems of information dissemination investigated in the current literature: *broadcasting*, *accumulation* and *gossiping*. For all three problems we may view any interconnection network as an undirected graph $G = (V, E)$, where the nodes in V correspond to the processors and the edges in E correspond to the communication links of the network. This abstraction is allowed because these three problems are purely communication problems, i. e. we do not need to deal with the computing actions of the processors.

Now, we make more precise what broadcast problem, accumulation problem and gossip problem mean.

1. **Broadcast problem for a graph G and a node v of G**

   Let $G = (V, E)$ be a graph and let $v \in V$ be a node of G. Let v know a piece of information $I(v)$ which is unknown to all nodes in $V \setminus \{v\}$. The problem is to find a communication strategy such that all nodes in G learn the piece of information $I(v)$.

2. **Accumulation problem for a graph G and a node v of G**

   Let $G = (V, E)$ be a graph, and let $v \in V$ be a node of G. Let each node $u \in V$ know a piece of information $I(u)$, and let, for any $x, y \in V$, the pieces of information $I(x)$ and $I(y)$ be "disjoined" (independent). The set $I(G) = \{I(w) \mid w \in V\}$ is called the *cumulative message* of G. The problem is to find a communication strategy such that the node v learns the cumulative message of G.

3. **Gossip problem for a graph G**

   Let $G = (V, E)$ be a graph, and let, for all $v \in V$, $I(v)$ be a piece of information residing in v. The problem is to find a communication strategy such that each node from V learns the whole cumulative message.

As we have seen above, all these communication problems are very natural for parallel networks. The broadcast problem is to spread the knowledge of one processor to all other processors in the network, the accumulation problem is to accumulate the knowledge of all processors in one given processor, and the gossip problem is to accumulate the knowledge of all processors in each processor of the network. Obviously, the description above provides only an abstract characterization of broadcasting, accumulating, and gossiping. To make the characterization more precise, we have to explain the notion "communication strategy". The communication strategy means for us a communication algorithm (also called *communication scheme*) from an allowed set of communication algorithms. Each communication algorithm is a sequence of simple communication steps called *communication rounds* (or simply *rounds*). To specify the set of allowed communication algorithms one defines a so-called communication mode which precisely describes what may happen in one communication round, i. e., the way how the edges (communication links) may be used or may not be used in one communication step. There are several communication modes investigated in the literature. Now, we present the one-way mode and the two-way mode which belong among the most extensively studied ones. The other modes will be shortly discussed in the last section.

a) **one-way mode** (also called **telegraph communication mode**)

In this mode, in a single round, each node may be active only via one of its adjacent edges either as sender or as receiver. It means that the information flow is one-way, i. e., one node sends a message to a given adjacent node. Thus, a one-way communication algorithm for a graph $G = (V, E)$ can be described as a sequence $E_1, E_2, \ldots, E_m$ of sets (matchings) $E_i \subseteq \vec{E}$, where $\vec{E} = \{(v \to u), (u \to v) \mid (u, v) \in E\}$ and if $(x_1 \to y_1), (x_2 \to y_2) \in E_i$ and $(x_1, y_1) \neq (x_2, y_2)$ for some $i \in \{1, \ldots, m\}$, then $x_1 \neq x_2 \wedge x_1 \neq y_2 \wedge y_1 \neq x_2 \wedge y_1 \neq y_2$.

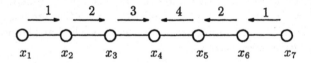

Figure 5.1

In Fig. 5.1 an accumulation algorithm for the path of 7 nodes and the node x_4 is depicted. In the first round the node x_1 sends its whole knowledge to x_2, and x_7 sends its knowledge to x_6. In the second round x_2 sends to x_3 and x_6 sends to x_5. In the third round x_3 sends to x_4, and in the 4th round x_5 sends to x_4. Obviously, this communication algorithm can be described as $\{(x_1 \to x_2), (x_7 \to x_6)\}, \{(x_2 \to x_3), (x_6 \to x_5)\}, \{(x_3 \to x_4)\}, \{(x_5 \to x_4)\}$, and everybody can see that the properties of the one-way mode are satisfied.

We note that we shall use several distinct ways to present communication algorithms in this paper. But each of these ways will provide for each communication round the exact information which edges are active (and in which direction they are active).

b) **two–way mode** (also called **telephone communication mode**)

In this mode, in a single round, each node may be active only via one of its adjacent edges and if it is active then it simultaneously sends a message and receives a message through the given, active edge (communication link). To say it in another way, if one edge is used for communication, the information flow is bidirectional. Thus, a two–way communication algorithm for a graph $G = (V, E)$ can be described as a sequence $E_1, E_2, \ldots, E_r$ of some sets (matchings) $E_i \subseteq E$, where for each $i \in \{1, \ldots, r\}, \forall (x_1, y_1), (x_2, y_2) \in E_i : \{x_1, y_1\} \neq \{x_2, y_2\}$ implies $x_1 \neq x_2 \wedge x_1 \neq y_2 \wedge y_1 \neq y_2 \wedge y_1 \neq x_2$.

Fig. 5.2 describes the following gossip algorithm

$$\{(x_1, x_3), (x_2, x_4)\}, \{(x_1, x_2), (x_3, x_4)\}$$

for the ring of four nodes.

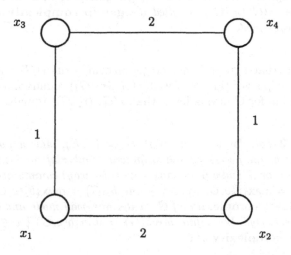

Figure 5.2

The last topic we have to explain is how the efficiency of communication algorithms is measured. We shall consider here one of the most used possibilities — the number of communication rounds. This measure does not deal with the length of the transmitted messages (with the amount of information exchanged). So, we assume that each node which is active as the sender in a given round sends its whole knowledge via the activated edge. The idea behind this is that one needs a lot of time to synchronize the network and to organize the information exchange in a given round, and the time needed for the direct information exchange via activated links is relatively small in comparison to the time for synchronization. This may be true in some cases, but there are also situations which require to measure also the time of any direct communication depending on the length of messages or the whole work of the network depending on the number of exchanged messages. These kinds of complexity measures will be discussed in the last section. Next, we shall only deal with the number of rounds which is one of the most commonly used complexity measures for communication algorithms.

Now, let us close this subsection by defining the complexity measures investigated.

Definition 5.1.1 *Let $G = (V, E)$ be a graph. Let $r(G)\,(r_2(G))$ denote the necessary and sufficient number of rounds for gossiping in G in the one–way (two–way) mode. $r(G)\,(r_2(G))$ is called the* **gossip complexity** *of G in the one–way (two–way) mode.*

Note that the fact that a graph G has the gossip complexity $r(G)$ $(r_2(G))$ means that there is a gossip algorithm for G with $r(G)$ $(r_2(G))$ rounds and there exists no gossip algorithm for G having fewer than $r(G)$ $(r_2(G))$ rounds.

Definition 5.1.2 *Let, for a given graph $G = (V, E)$, and a node $v \in V$, $b(v, G)\,(b_2(v, G))$ be the necessary and sufficient number of rounds to solve the broadcast problem for G and v in the one–way (two–way) communication mode. We define $b(G) = \max\{b(v, G) \mid v \in V\}$ and $b_2(G) = \max\{b_2(v, G) \mid v \in V\}$ to be the* **broadcast complexity** *of G in the one–way mode and in the two–way mode, respectively. We define $minb(G) = \min\{b(v, G) \mid v \in V\}$ as the* **min-broadcast complexity** *of G.*

Definition 5.1.3 *Let, for a given graph $G = (V, E)$, and a node $v \in V$, $a(v, G)\,(a_2(v, G))$ be the necessary and sufficient number of rounds to solve the accumulation problem for G and v in the one–way (two–way) mode. We define $a(G) = \max\{a(v, G) \mid v \in V\}$ and $a_2(G) = \max\{a_2(v, G) \mid v \in V\}$ to be the* **accumulation complexity** *of G in the one–way mode and in the two–way mode, respectively. We define $mina(G) = \min\{a_v(G) \mid v \in V\}$ as the* **min-accumulation complexity** *of G.*

5.1.2   Simple Observations and Relations between Complexity Measures

In this subsection we show that there is no difference between some of the complexity measures defined in the previous subsection, and so we show that it is sufficient to investigate only the broadcast problem in the one–way mode and the gossping problem in both modes. First, we shall show that we do not need the complexity measure defined in Definition 5.1.3 because the accumulation problem is exactly as hard as the broadcast problem for our modes. Note that this may be wrong for other communication modes. We shall mention a large difference between broadcast complexity and accumulation complexity for some other modes in the last section.

Observation 5.1.4 $a_2(v, G) = b_2(v, G)$ *for any graph G and any node v of G.*

Proof.   Let $E_1, E_2, \ldots, E_r$ be a broadcast algorithm for G and v in the two–way mode. Then $E_r, E_{r-1}, \ldots, E_2, E_1$ is an accumulation algorithm for G and v in the two–way mode. In the same way, a broadcasting scheme can be constructed from an accumulation algorithm needing the same number of rounds. $\square$

Corollary 5.1.5 $a_2(G) = b_2(G)$ *for any graph G.*

Observation 5.1.6 $a(v, G) = b(v, G)$ *for any graph G and any node v of G.*

Proof.   Let $E_1, E_2, \ldots, E_s$ be a broadcast algorithm for G and v in the one–way mode. Set $R_i = \{(x \to y) \mid (y \to x) \in E_i\}$. Then $R_s, R_{s-1}, \ldots, R_2, R_1$ is an accumulation algorithm for G and v in the one–way mode. In the same way, a broadcasting scheme can be constructed from an accumulation algorithm needing the same number of rounds. $\square$

Corollary 5.1.7 $a(G) = b(G)$ *and $mina(G) = minb(G)$ for any graph G.*

So, we see that it is sufficient to deal only with the broadcast complexity because all results for broadcast complexity hold also for accumulation complexity.

Now, we observe the intuitively clear fact that the two–way mode cannot help to decrease the broadcast complexity in comparison with the one–way mode, because for broadcasting it is sufficient that the information is flowing in one direction from the source node to all other nodes.

Observation 5.1.8 $b(v, G) = b_2(v, G)$ *for any graph* $G = (V, E)$ *and any node* v *of* G.

Proof.   It is clear from the definition that $b(v, G) \geq b_2(v, G)$, because the one–way mode cannot be more powerful than the two–way mode.

To prove $b(v, G) \leq b_2(v, G)$, let $A = E_1, E_2, \ldots, E_s$ be a broadcast algorithm for G and v in the two–way mode. Let R_i, $i = 1, \ldots, s$, be the set of nodes receiving the piece of information $I(v)$ in the first i rounds (i.e. during the run of the algorithm $E_1, E_2, \ldots, E_i$), and $R_0 = \{v\}$. Obviously, $\bigcup_{i=1}^{s} R_i = V$. Let $V_i = R_i \setminus \bigcup_{j=1}^{i-1} R_j$ (see Fig. 5.3). So, for $i = 0, 1, \ldots, s$, V_i is the set of nodes which receive $I(v)$ exactly in the i-th round and not before. Obviously $\bigcup_{i=0}^{s} V_i = V$ and $V_c \cap V_d = \emptyset$ for $c \neq d$, $c, d \in \{0, \ldots, s\}$.

Now, we remove the unnecessary edges (for example, (x, y) and (u, v) from E_4 in Fig 5.3) from the broadcast algorithm A in order to get the broadcast algorithm $A' = E_1', E_2', \ldots, E_s'$, where $E_i' = E_i \cap \left(\bigcup_{k=1}^{i-1} V_k \times V_i \right)$ for $i = 1, \ldots s$. Obviously A' is a broadcast algorithm in two–way mode with the property that each node from $V \setminus \{v\}$ receives $I(v)$ exactly once. So, the graphical representation of A' is a tree (see Fig. 5.3).

Now, to get a broadcast algorithm in one–way mode it is sufficient to direct the edges of A' in the direction from the root v to the leaves. Thus, $B = Z_1, Z_2, \ldots, Z_s$, where $Z_i = \{(x_1 \rightarrow x_2) \mid (x_1, x_2) \in E_i' \wedge x_1 \in \bigcup_{k=1}^{i-1} V_k \wedge x_2 \in V_i\}$ for $i = 1, \ldots, s$. It is clear that B is a communication algorithm in one–way mode, and everybody can easily prove by induction that for $i = 1, \ldots, s$ all nodes in $\bigcup_{j=0}^{i} V_j = R_i$ know $I(v)$ after the i-th round (after $Z_1, Z_2, \ldots, Z_i$). Thus, B is a broadcast algorithm in one–way mode with the same number of rounds as A.   □

Corollary 5.1.9 $b(G) = b_2(G)$ *for any graph* G.

The proof of Observation 5.1.8 shows that any broadcast algorithm of G and v determines a spanning tree of G rooted at v. Let us call this tree a **broadcast tree** of G and v.

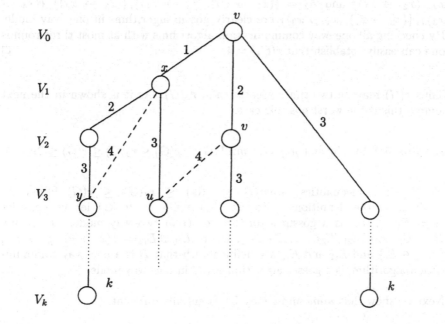

Figure 5.3

We see that it is sufficient to deal with the complexity measures r, r_2, b and $minb$ because all others are identical with one of these four. In what follows we shall show that these four measures are really different, and so we have to deal with all of them.

Example 5.1.10 Let us consider the ring C_4 of 4 nodes as depicted in Fig. 5.2. We see from Fig 5.2 that $r_2(C_4) = 2$. The following algorithms $A_1 = \{(x_1 \to x_3), (x_2 \to x_4)\}, \{(x_3 \to x_1), (x_4 \to x_2)\}\{(x_2 \to x_4), (x_1 \to x_2)\}, \{(x_4 \to x_2), (x_2 \to x_1)\}$ and $A_2 = \{(x_1 \to x_3), (x_2 \to x_4)\}, \{(x_3 \to x_4)\}, \{(x_4 \to x_3)\}, \{(x_3 \to x_1), (x_4 \to x_2)\}$ are clearly gossip algorithms in one–way mode. By checking all one–way communication algorithms with at most three rounds one can easily establish that $r(C_4) = 4$. $\qquad\square$

Thus $r(G)$ may be two times greater than $r_2(G)$. As it is shown in the next lemma this is the worst possible case.

Lemma 5.1.11 *For any graph G: $minb(G) \le b(G) \le r_2(G) \le r(G) \le 2r_2(G)$.*

Proof. The inequalities $minb(G) \le b(G) \le r_2(G) \le r(G)$ follow directly from the definitions. To see that $r(G) \le 2r_2(G)$ let us consider $A = E_1, \ldots, E_r$ as a gossip algorithm for G in two–way mode. Then any $B = E_{11}, E_{12}, E_{21}, E_{22}, \ldots, E_{r1}, E_{r2}$, where $E_{i1} \cup E_{i2} = \{(x \to y), (y \to x) \mid (x, y) \in E_i\}$ and E_{i1} and E_{i2} are defined such that B is a one–way communication algorithm, is a gossip algorithm for G in one–way mode. $\qquad\square$

Next we show that $minb$ and b may be essentially different.

Example 5.1.12 Let us consider a path P_n of n nodes $x_1, x_2, \ldots, x_n$ (see Fig 5.4). Obviously, $minb(P_n) = b(x_{\lceil n/2 \rceil}, P_n) = \lceil n/2 \rceil$, because

$$\{(x_{n/2} \to x_{n/2+1})\}, \{(x_{n/2} \to x_{n/2-1}), (x_{n/2+1}, x_{n/2+2})\},$$
$$\ldots, \{(x_3 \to x_2), (x_{n-2} \to x_{n-1})\}, \{(x_2 \to x_1), (x_{n-1} \to x_n)\}$$

is a broadcast algorithm for $P_n, x_{n/2}$ and n even , and

$$\{(x_{\lceil n/2 \rceil} \to x_{\lceil n/2 \rceil+1})\}, \{(x_{\lceil n/2 \rceil} \to x_{\lceil n/2 \rceil-1}), (x_{\lceil n/2 \rceil+1} \to x_{\lceil n/2 \rceil+2})\},$$
$$\ldots, \{(x_{n-1} \to x_n), (x_3 \to x_2)\}, \{(x_2 \to x_1)\}$$

is a broadcast algorithm for P_n, $x_{\lceil n/2 \rceil}$ and n odd. (No algorithm with fewer rounds than the two above exists because the distance between $x_{\lceil n/2 \rceil}$ and x_1

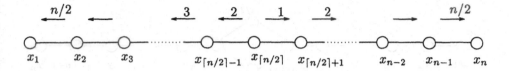

Figure 5.4 broadcasting in P_n for even n

is $\lceil n/2 \rceil - 1$, the distance between $x_{\lceil n/2 \rceil}$ and x_n is $\lfloor n/2 \rfloor$, and $x_{\lceil n/2 \rceil}$ can send $I(x_{\lceil n/2 \rceil})$ in the first round only in one direction). Clearly, $b(P_n) = b(x_1, P_n) = n - 1$ because the distance between x_1 and x_n is exactly $n - 1$. $\qquad \square$

So, we have seen that $b(G)$ may be almost two times as large as $minb(G)$. The next lemma shows that the difference cannot be greater.

Lemma 5.1.13 $b(G) \leq r(G) \leq 2 \cdot minb(G)$ *for any graph G of at least two nodes.*

Proof. Let $G = (V, E)$ be a graph, and let $v \in V$ be a node with the property $b(v, G) = minb(G)$. Let $A = E_1, E_2, \ldots, E_z$ for $z = minb(G)$ be a one–way broadcast algorithm for G and v. According to Observation 5.1.6 there exists a one–way accumulation algorithm $B = D_1, D_2, \ldots, D_z$ for G and v. Obviously, the concatenation of B and A: $B \circ A = D_1, D_2, \ldots, D_z, E_1, E_2, \ldots, E_z$ is a one–way gossip algorithm for G. So, $r(G) \leq 2 \cdot minb(G)$. $\qquad \square$

To see that there exist graphs for which the equality $r(G) = 2\, minb(G)$ holds it is sufficient to take the paths P_n for even n from Example 5.1.12. In this case, $minb(G) = n/2$ and $r(G) = n$. The latter fact will be proved in detail in the third section (Theorem 5.3.6). In that section, we shall also deal with the question for which other graphs the equality $r(G) = 2\, minb(G)$ holds.

A graph for which $b(G) = r(G)$ is the graph D_n as displayed in Fig 5.5 for even $n \geq 8$.

Clearly, $b(D_n) = n - 2$, because the distance between x_1 and x_n is $n - 2$ and broadcasting can be achieved in the same number of rounds. Also, $r(G) = n-2$, because

$$\{(x_1 \to x_2), (x_n \to x_{n-1}), (x_{n/2} \to x_{n/2-1}), (x_{n/2+1} \to x_{n/2+2})\},$$

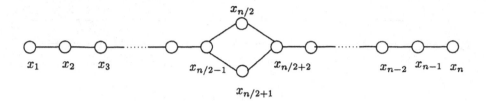

Figure 5.5 The graph D_n

$$\{(x_2 \to x_3), (x_{n-1} \to x_{n-2})\}, \ldots,$$
$$\{(x_{n/2-2} \to x_{n/2-1}), (x_{n/2+3} \to x_{n/2+2}\},$$
$$\{(x_{n/2-1} \to x_{n/2+1}), (x_{n/2+2} \to x_{n/2})\},$$
$$\{(x_{n/2} \to x_{n/2-1}), (x_{n/2+1} \to x_{n/2+2})\}, \ldots,$$
$$\{(x_3 \to x_2), (x_{n-2} \to x_{n-1}\}, \{(x_2 \to x_1), (x_{n-1} \to x_n)\}$$

is a gossip algorithm for D_n taking $n - 2$ rounds.

We close this subsection by showing how to get some straightforward lower bounds on the complexity of broadcasting and gossiping by investigating only some basic properties of graphs.

Definition 5.1.14 *Let $G = (V, E)$ be a graph, and let $u, v \in V$. The* **distance** *between u and v, $d(u, v)$, is the number of edges of the shortest path between u and v. The* **diameter** *$d(G)$ of G is the maximum distance between two nodes of G, i.e. $d(G) = \max\{d(u, v) \mid u, v \in V\}$. The* **radius** *of G is defined as $\mathrm{rad}(G) = \min_{v \in V} \max_{x \in V} d(v, x)$, the* **degree** *of G as $\deg(G) = \max_{v \in V} |\{(v, x) \in E\}|$.*

Observation 5.1.15

$$\mathrm{rad}(G) \leq minb(G) \text{ for any graph } G = (V, E).$$

Proof. Obviously, for each $v, x \in V$, $b(v, G)$ must be at least the distance $d(v, x)$.   □

The following observation follows also directly from the definitions.

Observation 5.1.16
$$\text{rad}(G) \le d(G) \le b(G).$$

The equalities $\text{rad}(G) = minb(G)$ and $d(G) = b(G)$ are satisfied again for the paths of odd length (i.e. with an even number of nodes).

In all what follows we shall denote by $V(G)$ the set of nodes of a given graph G, and by $E(G)$ the set of edges of G. If, for two graphs G_1 and G_2, $V(G_1) = V(G_2)$ and $E(G_1) \subseteq E(G_2)$ then we say that G_1 is a *spanning subgraph* of G_2. The following fact is obvious.

Observation 5.1.17 *For any* $x \in \{b, minb, r, r_2\}$ *and any two graphs* G_1 *and* G_2 *such that* G_1 *is a spanning subgraph of* G_2

$$x(G_2) \le x(G_1)$$

holds.

5.1.3 Definitions of Interconnection Networks

In this subsection, we provide the definitions of the most studied networks, and we fix their notation for the rest of the paper. For more information about these networks, we refer to [34].

The Path P_n. The *(simple) path of lenght n*, denoted by P_n, is the graph whose nodes are all integers from 1 to n and whose edges connect each integer i $(1 \le i < n)$ with $i + 1$.

P_n has n nodes, diameter $n - 1$ and maximum degree 2. An illustration of P_n is shown in Figure 5.6.

Figure 5.6 The path P_n

The Cycle C_n. The *cycle/ring of lenght n*, denoted by C_n, is the graph whose nodes are all integers from 1 to n and whose edges connect each integer i $(1 \leq i \leq n)$ with $i + 1(\bmod n)$.

C_n has n nodes, diameter $\lfloor n/2 \rfloor$ and maximum degree 2. An illustration of C_4 is shown in Figure 5.7.

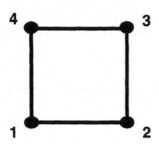

Figure 5.7   The cycle C_4

The Complete Tree $T_k{}^m$. The *complete k-ary tree of height m*, denoted by $T_k{}^m$, is the graph whose nodes are all k-ary strings of length at most m and whose edges connect each string α of length i $(0 \leq i \leq m)$ with the strings αa, $a \in \{0, \dots, k-1\}$, of length $i + 1$. The node ε, where ε is the empty string, is the *root* of $T_k{}^m$ and a node α is at *level i*, $i \geq 0$, in $T_k{}^m$ if α is a string of length i. The nodes at level m are the *leaves* of the tree. For a node α at level i, $0 \leq i < m$, the nodes αa, $a \in \{0, \dots, k-1\}$, are called the *sons/children* of α. α is called the *father/parent* of αa. For any node α, the nodes αu, $u \in \{0, \dots, k-1\}^*$, are called *descendants* of α, and α is called an *ancestor* of αu.

$T_k{}^m$ has $(k^{m+1} - 1)/(k - 1)$ nodes, diameter $2m$ and maximum degree $k + 1$. An illustration of $T_2{}^3$ is shown in Figure 5.8.

The Complete Graph K_n. The *complete graph/clique of size n*, denoted by K_n, is the graph whose nodes are all integers from 1 to n and whose edges connect each integer i, $1 \leq i \leq n$, with each integer j, $1 \leq i \leq n$, $j \neq i$.

K_n has n nodes, diameter 1 and maximum degree $n - 1$. An illustration of K_4 is shown in Figure 5.9.

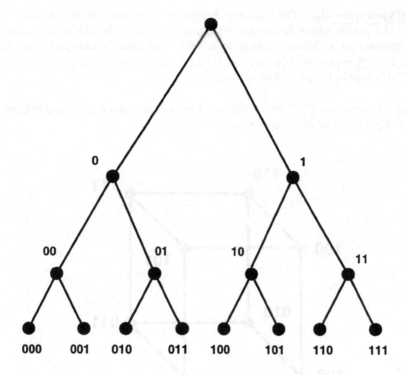

Figure 5.8 The complete tree $T_2{}^3$

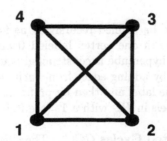

Figure 5.9 The complete graph K_4

The Hypercube H_m. The *(binary) hypercube* of dimension m, denoted by H_m, is the graph whose nodes are all binary strings of length m and whose edges connect those binary strings which differ in exactly one position. For each i, $1 \leq i \leq m$, an edge $(a_1 a_2 \ldots a_{i-1} 0 a_{i+1} \ldots a_m, a_1 a_2 \ldots a_{i-1} 1 a_{i+1} \ldots a_m)$, $a_l \in \{0,1\}$, is said to be in *dimension i*.

H_m has 2^m nodes, $m \cdot 2^{m-1}$ edges, diameter m and each node has exactly degree m. An illustration of H_3 is shown in Figure 5.10.

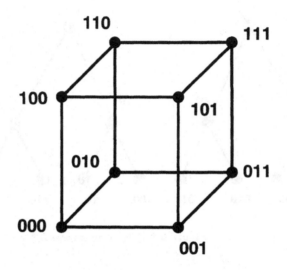

Figure 5.10   The hypercube H_3

Hypercubes may also be defined recursively as follows. A 1-dimensional hypercube is an edge with one vertex labeled 0 and the other labeled 1. An $(m + 1)$-dimensional hypercube is constructed from two m-dimensional hypercubes, H_m^0 and H_m^1, by adding edges from each vertex in H_m^0 to the vertex in H_m^1 that has the same label and then by prefixing all of the labels in H_m^0 with a 0 and all of the labels in H_m^1 with a 1. (see Figure 5.11).

The Cube-Connected Cycles CCC_m. The *cube-connected cycles network of dimension m*, denoted by CCC_m, has vertex-set $V_m = \{0, 1, ..., m-1\} \times \{0,1\}^m$, where $\{0,1\}^m$ denotes the set of length-m binary strings. For each vertex $v = \langle i, \alpha \rangle \in V_m$, $i \in \{0, 1, ..., m - 1\}, \alpha \in \{0,1\}^m$, we call i the level and α the position-within-level (PWL) string of v. The edges of CCC_m are of two types:

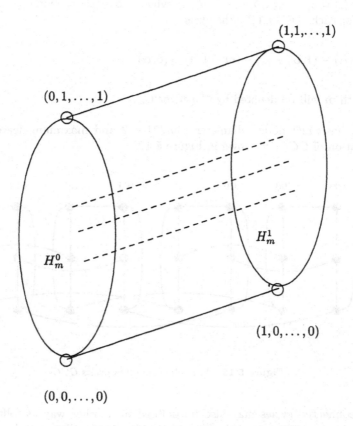

Figure 5.11 Constructing H_{m+1} from two H_m's

For each $i \in \{0, 1, ..., m-1\}$ and each $\alpha = a_0 a_1 ... a_{m-1} \in \{0, 1\}^m$, the vertex $\langle i, \alpha \rangle$ on level i of CCC_m is connected

- by a *straight-edge* with vertex $\langle i+1(\bmod m), \alpha \rangle$ on level $i+1(\bmod m)$ and

- by a *cross-edge* with vertex $\langle i, \alpha(i) \rangle$ on level i.

Here, $\alpha(i) = a_0 ... a_{i-1} \bar{a}_i a_{i+1} ... a_{m-1}$, where $\bar{a}$ denotes the binary complement of a. For each $\alpha \in \{0, 1\}^m$, the cycle

$$(0, \alpha) - (1, \alpha) - ... - (m-1, \alpha) - (0, \alpha)$$

of length m will be denoted by $C_\alpha(m)$ or C_α.

CCC_m has $m2^m$ nodes, diameter $\lfloor 5m/2 \rfloor - 2$ and maximum degree 3. An illustration of CCC_3 is shown in Figure 5.12.

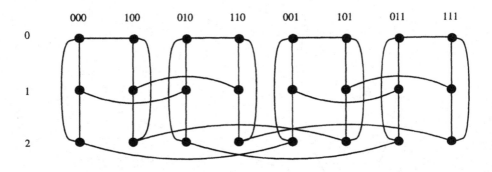

Figure 5.12 The cube-connected cycles CCC_3

Cube-connected cycles may also be defined in another way as follows. The CCC_m is a modification of the hypercube H_m obtained by replacing each vertex of the hypercube with a cycle of m processors. The i-th dimension edge incident to a node of the hypercube is then connected to the i-th node of the corresponding cycle of the CCC_m. For example, see Figure 5.13

The Butterfly BF_m. The *butterfly network of dimension m*, denoted by BF_m, has vertex-set $V_m = \{0, 1, ..., m-1\} \times \{0, 1\}^m$, where $\{0, 1\}^m$ denotes the set

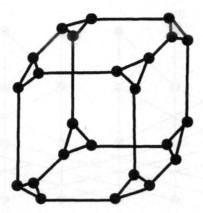

Figure 5.13   The cube-connected cycles CCC_3 as derived from H_3

of length-m binary strings. For each vertex $v = \langle i, \alpha \rangle \in V_m$, $i \in \{0, 1, ..., m - 1\}$, $\alpha \in \{0, 1\}^m$, we call i the level and α the position-within-level (PWL) string of v. The edges of BF_m are of two types: For each $i \in \{0, 1, ..., m - 1\}$ and each $\alpha = a_0 a_1 ... a_{m-1} \in \{0, 1\}^m$, the vertex $\langle i, \alpha \rangle$ on level i of BF_m is connected

-   by a *straight-edge* with vertex $\langle i + 1 (\mathrm{mod}\, m), \alpha \rangle$ and

-   by a *cross-edge* with vertex $\langle i + 1 (\mathrm{mod}\, m), \alpha(i) \rangle$

on level $i + 1 (\mathrm{mod}\, m)$. Again, $\alpha(i) = a_0 \ldots a_{i-1} \bar{a}_i a_{i+1} \ldots a_{m-1}$, where $\bar{a}$ denotes the binary complement of a. For each $\alpha \in \{0, 1\}^n$, the cycle

$$(0, \alpha) - (1, \alpha) - \ldots - (n - 1, \alpha) - (0, \alpha)$$

of length m will be denoted by $C_\alpha(k)$ or C_α.

BF_m has $m2^m$ nodes, diameter $\lfloor 3m/2 \rfloor$ and maximum degree 4. An illustration of BF_3 is shown in Figure 5.14. To obtain a clearer picture, level 0 has been replicated.

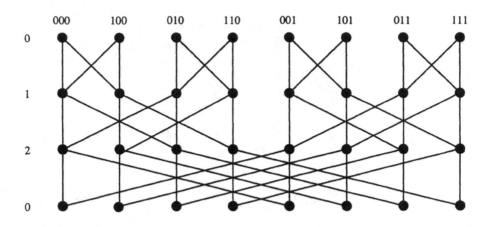

Figure 5.14   The butterfly graph BF_3

The Shuffle-Exchange SE_m. The *shuffle-exchange network of dimension* m, denoted by SE_m, is the graph whose nodes are all binary strings of length m and whose edges connect each string αa, where α is a binary string of length $m - 1$ and a is in $\{0, 1\}$, with the string $\alpha \bar{a}$ and with the string $a\alpha$. (An edge connecting αa with $\alpha \bar{a}$, is called an *exchange* edge and an edge connecting αa with $a\alpha$ is called a *shuffle* edge.)

SE_m has 2^m nodes, diameter $2m - 1$ and maximum degree 3. An illustration of SE_3 is shown in Figure 5.15.

The DeBruijn DB_m. The *deBruijn network of dimension* m, denoted by DB_m, is the graph whose nodes are all binary strings of length m and whose edges connect each string $a\alpha$, where α is a binary string of length $m - 1$ and a is in $\{0, 1\}$, with the strings αb, where b is a symbol in $\{0, 1\}$. (An edge connecting $a\alpha$ with αb, $a \neq b$, is called a *shuffle-exchange* and an edge connecting $a\alpha$ with αa is called a *shuffle* edge.)

DB_m has 2^m nodes, diameter m and maximum degree 4. An illustration of DB_3 is shown in Figure 5.16.

The Grid $[a_1 \times a_2 \times \ldots \times a_d]$. The *d-dimensional grid/mesh* of dimensions $a_1, a_2, \ldots, a_d$, denoted by $[a_1 \times a_2 \times \ldots \times a_d]$, is the graph whose nodes are

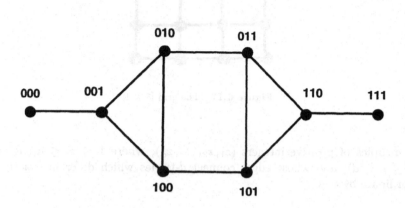

Figure 5.15   The shuffle-exchange graph SE_3

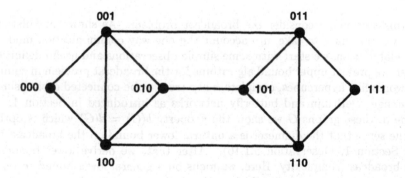

Figure 5.16   The deBruijn graph DB_3

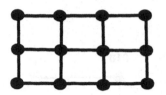

Figure 5.17   The grid $[3 \times 4]$

all d-tuples of positive integers $(z_1, z_2, \ldots, z_d)$, where $1 \leq z_i \leq a_i$, for all i $(0 \leq i \leq d)$, and whose edges connect d-tuples which differ in exactly one coordinate by one.

$[a_1 \times a_2 \times \ldots \times a_d]$ has $a_1 \times a_2 \times \ldots \times a_d$ nodes, diameter $(a_1 - 1) + (a_2 - 1) + \ldots + (a_d - 1)$ and maximum degree $2d$, if each a_i is at least three. An illustration of $[3 \times 4]$ is shown in Figure 5.17.

5.2   BROADCASTING

5.2.1   Introduction

In this section we consider the broadcast problem. As shown in Subsection 1.2, we only have to take into account the one-way communication mode. In our elaboration, we start with some simple observations and useful definitions. Then we present upper bound algorithms for the broadcast problem in common networks like hypercubes, complete k-ary trees, cube-connected cycles, shuffle-exchange, DeBruijn and butterfly networks as introduced in Section 1. For some of these graphs G we show the property $b(G) = d(G)$ which is optimal in the sense that the diameter is a natural lower bound on the broadcast time (cf. Section 1, Observation 5.1.16). After that, we derive lower bounds on the broadcast complexity. Here, we focus on a general lower bound technique for bounded–degree graphs which we apply in every special case. Finally, we discuss some related results and open problems, and we give an overview of the presented results.

Before going into more sophisticated results, let us start off with a simple general lower bound for the broadcast problem:

Observation 5.2.1 *Let G be a graph with n nodes. Then*

$$b(G) \geq minb(G) \geq \lceil \log_2 n \rceil.$$

Proof. $b(G) \geq minb(G)$ is clear. To prove $minb(G) \geq \lceil \log_2 n \rceil$, let $A(t)$ denote the maximum number of nodes which can know the message after t rounds. As the number of informed nodes can at most double during each time unit, we have the following recursive definition:

$$A(0) = 1,$$
$$A(t + 1) = 2 \cdot A(t) \quad \text{for all } t \geq 0.$$

It is easy to verify that the closed formula for $A(t)$ is

$$A(t) = 2^t.$$

Therefore, at most 2^t nodes are informed after t rounds. To inform all n nodes, the relation

$$2^t \geq n$$

must hold, hence $t \geq \lceil \log_2 n \rceil$. □

Now, the first question arising is whether there are graphs of n nodes satisfying the property $b(G) = \lceil \log_2 n \rceil$. These graphs are called **minimal broadcast graphs**. We show that the complete graph and any graph having the hypercube as a subgraph has this property. The crucial point is that it must be possible to double the number of informed nodes in each round.

Lemma 5.2.2

   a) $b(K_n) = \lceil \log_2 n \rceil$,

   b) $b(H_m) = m$.

Proof.

a) Number the nodes of K_n from 0 to $n - 1$. W.l.o.g. let the originating node be 0. The following algorithm has the property that it doubles the number of informed nodes in each round.

Algorithm BROADCAST-K_n
<u>for</u> $t = 1$ <u>to</u> $\lceil \log_2 n \rceil$ <u>do</u>
   <u>for</u> all $i \in \{0, \ldots, 2^{t-1} - 1\}$ <u>do</u> in <u>parallel</u>
      <u>if</u> $i + 2^{t-1} \leq n$ <u>then</u>
         i sends to $i + 2^{t-1}$;

It is easy to verify by induction on t that after t rounds of the algorithm, the nodes $0, 1, \ldots, \min\{2^t - 1, n\}$ have been informed. Therefore, after $\lceil \log_2 n \rceil$ rounds, all nodes have received the information.

b) The algorithm for the hypercube H_m is exactly the same as for the complete graph K_n, where $n = 2^m$. Using the binary representation of the nodes, w.l.o.g. the originating node is $00 \ldots 0$, and algorithm BROADCAST-K_n directly translates into

Algorithm BROADCAST-H_m
<u>for</u> $i = 1$ <u>to</u> m <u>do</u>
   <u>for</u> all $a_0, \ldots, a_{i-1} \in \{0, 1\}$ <u>do</u> in <u>parallel</u>
      $a_0 \ldots a_{i-1}00 \ldots 0$ sends to $a_0 \ldots a_{i-1}10 \ldots 0$;

In other words, in round i, each informed vertex sends the message in dimension i ($1 \leq i \leq m$). From part a), we know that after m rounds all the nodes have received the information.

$\square$

We note that there is no bounded–degree interconnection network G of n nodes having the property $b(G) = \lceil \log_2 n \rceil$, because the doubling of the informed nodes is only possible if each node is active in each round via another edge. A detailed analysis of this situation is presented in Subsection 2.3.

Another interesting problem is to find graphs having the property $b(G) = \lceil \log_2 n \rceil$ and as few edges as possible. These graphs are called **minimum broadcast graphs**. This question has been investigated in several papers (for an overview, see e.g. [22, 18]). Due to lack of space, we do not consider this problem here.

From Section 1 (Observation 5.1.16) we know that

$$d(G) \leq b(G) \quad \text{for all graphs } G,$$

i.e. the diameter of the graph is a trivial lower bound on the broadcast time. As $d(H_m) = m$, Lemma 5.2.2 shows that the hypercube H_m is another example for a graph G satisfying $d(G) = b(G)$. In Subsection 2.2 we will find other graphs which are optimal (or near optimal) in this sense.

The diameter lower bound can be slightly improved in many cases as follows:

Observation 5.2.3 *Let G be a graph of diameter D. If there exist three different vertices u, v_1 and v_2 with both v_1 and v_2 at distance D from u, then*

$$b(G) \geq D + 1.$$

Proof. Let S be a broadcasting scheme for G and v. By induction on i, we can see that in round i of the scheme, at most one vertex at distance i from the originator v can be informed. Therefore, to inform two nodes v_1 and v_2 at distance D, at least $D + 1$ rounds are needed. $\square$

This observation will turn out to be quite useful for the cube-connected-cycles network in Subsection 2.2. A generalization of the idea contained in the proof of Observation 5.2.3 will lead to more powerful lower bounds on the broadcast time in Subsection 2.3.

We conclude this subsection by presenting two elementary but very instructive examples for broadcasting in certain types of networks. First, let us recall from Section 1 that any broadcast algorithm of a graph G and a node v determines a spanning tree of G rooted at v. This tree is called a **broadcast tree** of G and v. It turns out that this description of broadcast algorithms is quite useful for proving lower bounds on the broadcast time. This is demonstrated in the following instructive example determining the min-broadcast time of the k-ary tree:

Lemma 5.2.4
$$minb(T_k{}^m) = k \cdot m.$$

Proof. Let v_0 be the root of $T_k{}^m$. We show that

1. $b(v_0, T_k{}^m) = k \cdot m$,

2. $b(v_0, T_k{}^m) \leq b(v, T_k{}^m)$ for all $v \in V(T_k{}^m)$.

Statements 1. and 2. imply the validity of the lemma. —

1. $\underline{b(v_0, T_k{}^m) = k \cdot m}$:

   First, we show that it is possible to broadcast from v_0 in $k \cdot m$ rounds. The algorithm works as follows:

   Algorithm MINBROADCAST-$T_k{}^m$

   1. The root v_0 learns the message at time 0.
   2. Each non-leaf node receiving the message at time t, sends it on to its k sons in the next k rounds.

   It is straightforward to see that after $k \cdot i$ rounds ($1 \leq i \leq m$), each node at distance at most i from the root has received the information. Hence, after $k \cdot m$ rounds, each node in the tree has received the information. —

   Now, we show that any broadcast from v_0 takes at least $k \cdot m$ rounds. Let T be a broadcast tree of $T_k{}^m$ rooted at v_0. Label the edges of T as follows:

   Let v be any non-leaf vertex of T and $v_1, \ldots, v_k$ be the sons of v. Suppose that vertex v receives the message at time t and vertex v_i receives the message at time $t + i$ from v ($i = 1, \ldots, k$). Then we label the edge connecting v with v_i by i ($i = 1, \ldots, k$). See Fig. 5.18.

   As $T_k{}^m$ is a complete tree, it is clear that there is a path of length m in T from the root v_0 to some leaf w which is only labeled with k's, and w is informed via this path. Hence, w receives the message at time $k \cdot m$.

2. $\underline{b(v_0, T_k{}^m) \leq b(v, T_k{}^m)}$ for all $v \in V(T_k{}^m)$:
   Number the subtrees rooted at the sons of v_0 from $T_1, \ldots, T_k$ (see Fig. 5.19). W.l.o.g. let the originator v of the message be a node of the subtree T_1. To inform the subtrees $T_2, \ldots, T_k$, the message has to pass through the root v_0. Once v_0 learns the message, it has to be broadcast in the whole tree $T_k{}^m$ except for the subtree T_1. This subtree $\tilde{T}$ is shown in Fig. 5.20.

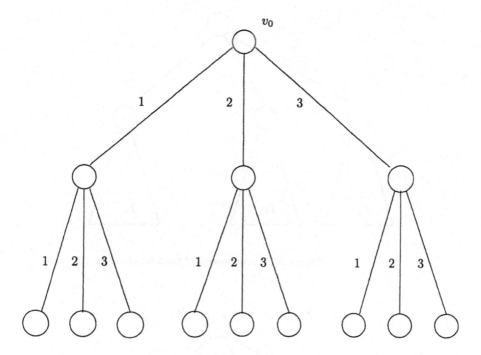

Figure 5.18   Labeling of edges of T

Using exactly the same arguments as for the lower bound in part 1., it can be shown that broadcasting in $\tilde{T}$ takes at least $k \cdot m - 1$ rounds. As at least one round is needed to inform v_0, we have

$$b(v, T_k{}^m) \geq k \cdot m = b(v_0, T_k{}^m).$$

$\square$

More sophisticated lower bound techniques will be presented in Subsection 2.3. There we look at graphs of bounded degree. A simple upper bound on the broadcast time of these graphs can be obtained as follows:

Lemma 5.2.5 *Let G be a graph of degree d. Then*

    a) $minb(G) \leq (d-1) \cdot \mathrm{rad}(G) + 1,$

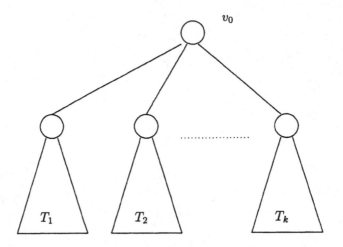

Figure 5.19 Numbering of the subtrees of v_0

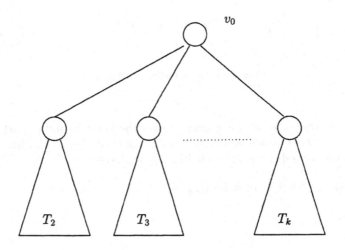

Figure 5.20 The subtree $\tilde{T}$

b) $b(G) \leq (d-1) \cdot d(G) + 1.$

Proof.

a) Let v_0 be a node in G such that $d(v,x) \leq \text{rad}(G)$ for all $x \in V$. The following algorithm broadcasts from any node w:

Algorithm MINBROADCAST-G

1. The root w learns the message at time 0.

2. w sends the message to its (at most) d uninformed adjacent vertices in the first d rounds.

3. Each node $v \neq w$ receiving the message at time t, sends it on to its (at most) $d-1$ still uninformed sons in the next $d-1$ rounds.

It is easy to see that after $(d-1) \cdot i + 1$ rounds, each node at distance at most i from w has received the information. Hence, when applying algorithm MINBROADCAST-G to the root v_0, after $(d-1) \cdot \text{rad}(G) + 1$ rounds, all the nodes in G have received the information. Thus,

$$b(v_0, G) \leq (d-1) \cdot \text{rad}(G) + 1,$$

and we have shown that

$$minb(G) \leq (d-1) \cdot \text{rad}(G) + 1.$$

b) Let w be any node in G. We apply algorithm MINBROADCAST-G from part a) to broadcast from w in time at most

$$(d-1) \cdot d(G) + 1.$$

Hence, we have that

$$b(G) \leq (d-1) \cdot d(G) + 1.$$

$\square$

We will see in Subsection 2.2 that the above stated upper bounds for broadcasting in bounded-degree graphs are not very sharp in general. But there are in fact cases in which this simple algorithm already yields the best possible result:

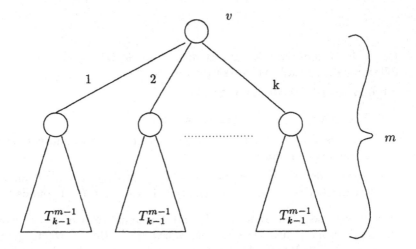

Figure 5.21 The tree $\tilde{T}_k{}^m$

Example 5.2.6 Let $\tilde{T}_k{}^m$ denote the k-ary tree consisting of a root v with k complete $(k-1)$-ary trees of depth $m-1$ as its sons (see Fig. 5.21). With the same techniques as used in the proof of Lemma 5.2.4 it can be shown that

$$minb(\tilde{T}_k{}^m) = (k-1) \cdot m + 1 = (\deg(\tilde{T}_k{}^m) - 1) \cdot \text{rad}(\tilde{T}_k{}^m) + 1 .$$

$\square$

5.2.2 Upper Bounds for Common Networks

In this subsection, we present upper bounds for broadcasting in popular networks, namely cube-connected cycles, butterfly, shuffle-exchange and DeBruijn networks. A summary of results for these networks can be found in [18]. The simple upper bounds for bounded-degree graphs from Lemma 5.2.5 yield (roughly)

1. $b(CCC_k) \leq 5k$,

2. $b(BF_k) \leq 4.5k$,

3. $b(SE_k) \leq 4k$,

4. $b(DB_k) \leq 3k$.

We will see that we can do a lot better than that in all cases. We start by looking at the cube-connected cycles network:

Theorem 5.2.7 ([32])
$$\lceil \frac{5k}{2} \rceil - 2 \leq b(CCC_k) \leq \lceil \frac{5k}{2} \rceil - 1.$$

Proof.

1. First, we prove that

$$b(CCC_k) \geq \left\lceil \frac{5k}{2} \right\rceil - 2.$$

To verify this, note that CCC_k has diameter $\lfloor 5k/2 \rfloor - 2$. If k is even, this implies that

$$b(CCC_k) \geq \left\lfloor \frac{5k}{2} \right\rfloor - 2 = \left\lceil \frac{5k}{2} \right\rceil - 2.$$

Let k be odd. W.l.o.g. the message originates at vertex $u = \langle 0, 00...0 \rangle$. There exist two nodes (namely $v_1 = \langle \lfloor k/2 \rfloor, 11...1 \rangle$ and $v_2 = \langle \lfloor k/2 \rfloor - 1, 11...1 \rangle$ at distance $\lfloor 5k/2 \rfloor - 1$ of u. From Observation 5.2.3, we obtain

$$b(CCC_k) \geq \left(\left\lfloor \frac{5k}{2} \right\rfloor - 2 \right) + 1 = \left\lceil \frac{5k}{2} \right\rceil - 2.$$

2. Now, we present an algorithm which broadcasts in time $\lceil 5k/2 \rceil - 1$ from $v_0 = \langle 0, 00...0 \rangle$:

Algorithm BROADCAST-CCC_k
1. $\langle 0, 00...0 \rangle$ sends to $\langle 0, 10...0 \rangle$;
   for $i = 1$ to $k - 1$ do
      begin for all $a_0, \ldots, a_{i-1} \in \{0, 1\}$ do in parallel
                $\langle i, a_0 \ldots a_{i-1}00 \ldots 0 \rangle$ sends to $\langle i+1, a_0 \ldots a_{i-1}00 \ldots 0 \rangle$;
             for all $a_0, \ldots, a_{i-1} \in \{0, 1\}$ do in parallel
                $\langle i, a_0 \ldots a_{i-1}00 \ldots 0 \rangle$ sends to $\langle i, a_0 \ldots a_{i-1}10 \ldots 0 \rangle$;
      end;

2. <u>for</u> all $\alpha \in \{0,1\}^k$ <u>do</u> in <u>parallel</u>
    broadcast on the cycle $C_\alpha(k)$ from $\langle k-1, \alpha \rangle$;

It can easily be verified by induction on i that after $2i-1$ rounds of Phase 1 ($1 \le i \le k$), the nodes $\langle i-1, a_0 \ldots a_{i-1} 00 \ldots 0 \rangle$, $a_0, \ldots, a_{i-1} \in \{0,1\}$ have received the information. Hence, after $2k-1$ rounds, i.e. after Phase 1, all the nodes $\langle k-1, \alpha \rangle$, $\alpha \in \{0,1\}^k$ have received the message. In Phase 2, broadcasting on the cycles $C_\alpha(k)$, $\alpha \in \{0,1\}^k$ can be done in $\lceil k/2 \rceil$ rounds (cf. Example 5.1.12). So, overall the algorithm takes $\lceil 5k/2 \rceil - 1$ rounds.

$\square$

Next, we investigate the shuffle-exchange network:

Theorem 5.2.8 ([24])
    $2k - 1 \le b(SE_k) \le 2k$.

Proof. The lower bound comes from the fact that SE_k has diameter $2k-1$. For the upper bound, let for each word $w = a_1 a_2 \ldots a_k \in \{0,1\}^k$, $w_1 = a_1$ and $w^t = a_{t+1} a_{t+2} \ldots a_k$ for $t \le k$. If $w = \epsilon$ then $w_1 = \epsilon$. Now, we shall write the broadcasting algorithm for an arbitrary source node α in SE_k.

Algorithm BROADCAST-SE_k
    <u>for</u> $t = 0$ <u>to</u> $k-1$ <u>do</u>
        <u>for</u> all $\beta \in \{0,1\}^t$ <u>do</u> in <u>parallel</u>
            <u>begin</u> <u>if</u> $\alpha^t \notin \{\beta_1\}^*$ <u>then</u>
                    <u>begin</u> $\alpha^t \beta$ sends to $\alpha^{t+1} \beta \alpha_1^t$ (shuffle round) <u>end</u>;
                $\alpha^{t+1} \beta \alpha_1^t$ sends to $\alpha^{t+1} \beta \overline{\alpha}_1^t$ (exchange round)
      <u>end</u>;

Now, we need to prove the following two facts:

(1) there is no conflict in any of $2k$ rounds, i.e. algorithm BROADCAST-SE_k works in the one-way mode (if a node is active in a round then it is active only via one edge in one direction).

(2) after $2r$ rounds (r executions of the loop) all nodes $\alpha^r \beta, \beta \in \{0,1\}^r$, have learned the piece of information of α.

(1) There is no conflict in any exchange round because each sender has the last bit α_1^t and each receiver has the last bit $\overline{\alpha}_1^t$. Let there be a conflict in a shuffle round, i.e., $\alpha^t \beta = \alpha^{t+1} \gamma \alpha_1^t$ for some $\beta, \gamma \in \{0,1\}^+$. It implies $\alpha_1^t \alpha^{t+1} = \alpha^{t+1} \gamma_1 \Rightarrow \alpha_1^t = \alpha_1^{t+1} = \ldots = \alpha_1^{k-1} = \gamma_1 \Rightarrow \alpha^t \in \{\gamma_1\}^*$. But this is a contradiction because we do not use Shuffle-operation for $\alpha^t \in \{\beta_1\}^*$.

(2) This will be proved by induction according to $r = t+1$. It has to be shown that the nodes $\alpha^{t+1} \beta \alpha_1^t, \alpha^t \in \{\beta_1\}^*$ (which do not receive the information in the r-th execution of the loop), have got the information already in previous rounds.

Clearly, our induction hypothesis [that all nodes $\alpha^r \beta$, for each $\beta \in \{0,1\}^r$ have learned the piece of information of α after r executions of the loop] is fulfilled after the first execution of the loop.

Now, let us consider the situation after r executions of the loop. Clearly, if $\alpha^{r-1} \notin \{\beta_1\}^*$ then all $\alpha^r \beta$ for $\beta \in \{0,1\}^r$ know the piece of information of α.
If $\alpha^{r-1} \in \{\beta_1\}^*$ then $\alpha^r \beta \alpha_1^r = \alpha^{r-1} \beta^1 \alpha_1^r$ which knows already the piece of information of α according to the induction hypothesis.

$\square$

The two previous results show that the upper bound algorithm for CCC_k and SE_k almost match the diameter lower bound. It turns out that this is not true for BF_k and DB_k.

Let us consider the butterfly network BF_k first. As the search for the best upper bound is still going on, we present an instructive yet very efficient algorithm by E. Stöhr [39] which needs $2k$ rounds for BF_k. This bound has been improved to $2k - 1$ by Klasing, Peine, Monien and Stöhr [28]. Refinements of these techniques [41] show that an upper bound of $2k - \frac{1}{2} \log \log k + c$, for some constant c and all sufficiently large k, is also possible. But for the sake of instructiveness, we have chosen to present the upper bound of $2k$. As for the lower bound, we first state the diameter lower bound. In Subsection 2.3, we will derive a non-trivial lower bound for broadcasting in BF_k.

Theorem 5.2.9 ([39])
$$\left\lfloor \frac{3m}{2} \right\rfloor \leq b(BF_m) \leq 2m.$$

Proof. The lower bound comes from the fact that BF_m has diameter $\lfloor 3m/2 \rfloor$. For the upper bound, first note that BF_m contains two isomorphic subgraphs F_0 and F_1. The subgraph F_0 has vertex set $\{\langle l\,;\,\alpha 0 \rangle \mid 0 \leq l \leq m-1,\ \alpha \in \{0,1\}^{m-1}\}$, and the subgraph F_1 has vertex set $\{\langle l\,;\,\alpha 1 \rangle \mid 0 \leq l \leq m-1,\ \alpha \in \{0,1\}^{m-1}\}$. Obviously, $F_0 \cap F_1 = \emptyset$.

Then note that BF_m contains 2^m node-disjoint cycles $C_{\alpha i}$ of the length m, $\alpha \in \{0,1\}^{m-1}$, $i \in \{0,1\}$, of the form $(\langle 0\,;\,\alpha i \rangle, \langle 1\,;\,\alpha i \rangle, \ldots, \langle m-1\,;\,\alpha i \rangle, \langle 0\,;\,\alpha i \rangle)$.

Let $\alpha \in \{0,1\}^{m-1}$ be any string of length $m-1$. By $\sharp_1(\alpha)$ we denote the number of 1's in α and by $\sharp_0(\alpha)$ we denote the number of 0's in α. So from the definition we have $\sharp_1(\alpha) + \sharp_0(\alpha) = m-1$.

Consider the node $v_0 = \langle 0;0\ldots0 \rangle$ of F_0. For every node $w_0 = \langle m-1\,;\,\alpha 0 \rangle$ of F_0, $\alpha \in \{0,1\}^{m-1}$, there is a path in F_0 of length $m-1$ connecting v_0 and w_0. This path can easily be constructed as follows: the path traverses the straight edge between level i and level $i+1$ for every bit position i in which α has a 0, and it traverses the cross edge between level i and level $i+1$ for every bit position i in which α has a 1, $0 \leq i \leq m-2$.

Now consider $v_1 = \langle m-1\,;\,0\ldots01 \rangle$, the level $m-1$ node of F_1. Similarly, there is a path in F_1 of length $m-1$ connecting v_1 with any level-0 node $w_1 = \langle 0\,;\,\alpha 1 \rangle$.

Since the butterfly network is a Cayley graph [2], and every Cayley graph is vertex symmetric [1], we can assume that the message originates at vertex $v_0 = \langle 0;0\ldots0 \rangle$, and the originator learns the message at time 0.

In the first step the node v_0 informs the neighbor $v_1 = \langle m-1;0\ldots01 \rangle$. Now as well in F_0 as in F_1 one node is informed. Then broadcasting in F_0 and F_1 will be done as follows in two phases:

Phase 1: In each cycle $C_{\alpha 0}$ inform the node $w_0 = \langle m-1\,;\,\alpha 0 \rangle$ and in each cycle $C_{\alpha 1}$ inform the node $w_1 = \langle 0\,;\,\alpha 1 \rangle$ in at most $\lfloor 3m/2 \rfloor$ rounds, $\alpha \in \{0,1\}^{m-1}$

The broadcasting scheme we use is a little different in F_0 and F_1. In F_0 we prefer the straight edges. This means, that any node $\langle l\,;\,\alpha 0 \rangle$, $0 \leq l \leq m-2$, $\alpha \in \{0,1\}^{m-1}$, of F_0 that receives the message at time t, informs its neighbor $\langle l+1\,;\,\alpha 0 \rangle$ at time $t+1$ and its neighbor $\langle l+1\,;\,\alpha(l)0 \rangle$ at time $t+2$.

In F_1 we prefer the cross–edges: from any node $\langle l \, ; \alpha 1 \rangle$, $1 \leq l \leq m - 1$, $\alpha \in \{0,1\}^{m-1}$, of F_1 that receives the message at time t, the neighbor $\langle l-1; \alpha(l)1 \rangle$ receives the message at time $t+1$ and the neighbor $\langle l-1; \alpha 1 \rangle$ receives the message at time $t + 2$.

Consider now any node $w_0 = \langle m - 1 \, ; \alpha 0 \rangle$, $\alpha \in \{0,1\}^{m-1}$, in F_0. The node gets the information from v_0 by broadcasting along the path in F_0 described above. This path traverses $\#_1(\alpha)$ cross edges and $\#_0(\alpha)$ straight edges. Since in F_0 the straight edges are preferred, w_0 is informed at time $1 + 2\,\#_1(\alpha) + \#_0(\alpha) = m + \#_1(\alpha)$. (You have to add 1 since in the first round the node v_0 informs v_1.)

Similarly, by using the path in F_1 described above, for all $\alpha \in \{0,1\}^{m-1}$ the node $w_1 = \langle 0 \, ; \alpha 1 \rangle$ is informed at time $1 + \#_1(\alpha) + 2\,\#_0(\alpha) = m + \#_0(\alpha)$.

Obviously for some $\alpha \in \{0,1\}^{m-1}$ the node w_0 or w_1 is informed in more than $\lfloor \frac{3m}{2} \rfloor$ rounds. For example for $m = 3$ the node $\langle 0; 001 \rangle$ is informed in F_1 in round $1 + \#_1(00) + 2\#_0(00) = 5 > 4 = \lfloor \frac{3m}{2} \rfloor$.

In these cases we inform the nodes by using the cross edges from level $m-1$ in F_0 to level 0 in F_1 or vice versa. In the example $m = 3$ we can inform the node $\langle 0; 001 \rangle$ by using the cross edge from $\langle 3; 000 \rangle$ to $\langle 0; 001 \rangle$. Since $\langle 3; 000 \rangle$ is informed in round $1 + \#_0(00) + 2\#_1(00) = 3$, the node $\langle 0; 001 \rangle$ is informed in round $4 = \lfloor \frac{3m}{2} \rfloor$.

In general, we consider the following cases:

Case 1: m odd

    Case 1.1 : $\#_1(\alpha) < (m - 1)/2$

    The node w_0 is informed from v_0 at time $m + \#_1(\alpha) < (3m - 1)/2 = \lfloor 3m/2 \rfloor$.

    In the next round, w_0 sends the message to its neighbor w_1. So w_1 is informed at time at most $\lfloor 3m/2 \rfloor$.

    Case 1.2: $\#_0(\alpha) < (m - 1)/2$

    The node w_1 is informed from v_0 at time $m + \#_0(\alpha) < (3m - 1)/2 = \lfloor 3m/2 \rfloor$.

    So the node w_0 which is adjacent to the node w_1 is informed at time at most $\lfloor 3m/2 \rfloor$.

    Case 1.3: $\#_0(\alpha) = \#_1(\alpha) = (m - 1)/2$

    w_0 is informed at time $m + \#_1(\alpha) = (3m - 1)/2 = \lfloor 3m/2 \rfloor$.

    w_1 is informed at time $m + \#_0(\alpha) = (3m - 1)/2 = \lfloor 3m/2 \rfloor$.

Case 2: m even

    Case 2.1: $\#_1(\alpha) \leq (m - 2)/2$

The node w_0 is informed from v_0 at time

$m + \sharp_1(\alpha) \le 3m/2 - 1 < \lfloor 3m/2 \rfloor$.

So the node w_1 is informed at time at most $\lfloor 3m/2 \rfloor$.

Case 2.2: $\sharp_0(\alpha) \le (m - 2)/2$

The node w_1 is informed from v_0 at time $m + \sharp_0(\alpha) \le 3m/2 - 1 < \lfloor 3m/2 \rfloor$.

So the node w_0 is informed at time at most $\lfloor 3m/2 \rfloor$.

So after phase 1, for all $\alpha \in \{0,1\}^{m-1}$ the nodes w_0 and w_1 received the message in at most $\lfloor 3m/2 \rfloor$ rounds.

Phase 2: Inform all nodes in the cycles $C_{\alpha i}$, $\alpha \in \{0,1\}^{m-1}$, $i \in \{0,1\}$

From the informed node we can inform all other nodes of the cycle in $\lceil m/2 \rceil$ rounds (cf. Example 5.1.12). So the broadcast time in the Butterfly network is at most $\lfloor 3m/2 \rfloor + \lceil m/2 \rceil = 2m$.

$\square$

Now we investigate the DeBruijn network DB_k. The best known upper bound so far was found by Bermond and Peyrat [7]. For the lower bound, we again state the diameter lower bound and refer to Subsection 2.3 for a non-trivial lower bound for broadcasting in DB_k.

Theorem 5.2.10 ([7])
$$d \le b(DB_d) \le \frac{3}{2}(d+1).$$

Proof. The lower bound comes from the fact that DB_d has diameter d. For the upper bound, the idea of the broadcasting scheme is that any node broadcasts only to its right neighbors (i.e. $(y_1, y_2, \ldots, y_d)$ informs its neighbors $(y_2, \ldots, y_d, y_1)$ and $(y_2, \ldots, y_d, \overline{y_1})$). The order of broadcasting will be determined according to the 2–arity α of $(y_1, y_2, \ldots, y_d)$, that is $\alpha(y_1, \ldots, y_d) = \left(\sum_{i=1}^{d} y_i \right) \bmod 2$. Note that $\alpha \in \{0,1\}$. The node $(y_1, y_2, \ldots, y_d)$ will broadcast to its right neigbors in the order $(y_2, \ldots, y_d, \alpha), (y_2, \ldots, y_d, \overline{\alpha})$. (e.g. in DB_6 for the node $(0,0,1,1,0,0)$ the value of α is 0, and so the node informs at first $(0,1,1,0,0,0)$ and then $(0,1,1,0,0,1)$). Now, consider the following two paths P_k, $k \in \{0,1\}$, of length $d+1$ from $(y_1, \ldots, y_d)$ to any node $(z_1, \ldots, z_d)$:

$$P_k \colon \big((y_1, \ldots, y_d), (y_2, \ldots, y_d, k)), (y_3, \ldots, y_d, k, z_1), (y_4, \ldots, y_d, k, z_1, z_2),$$

$$\dots (y_d, k, z_1, \dots, z_{d-2}), (k, z_1, \dots, z_{d-1}), (z_1, \dots, z_{d-1}, z_d)\Big).$$

Obviously, the paths are node-disjoint besides the first and the last node. Let $v_{0_i} = (y_i, \dots, y_d, 0, z_1, \dots, z_{i-2})$ $(v_{i_1} = (y_i, \dots, y_d, 1, z_1, \dots, z_{i-2}))$ be the i-th node of P_0 (P_1), $1 < i < d+2$. The nodes v_{0_i} and v_{1_i} differ just in one bitposition. So we have $\alpha(y_i, \dots, y_d, 0, z_1, \dots, z_{i-2}) = \alpha(y_i, \dots, \overline{y_d}, 1, z_1, \dots, z_{i-2}) \in \{0, 1\}$.

That means that the number of time units required to broadcast from $(y_i, \dots, y_d, 0, z_1, \dots, z_{i-2})$ to $(y_{i+1}, \dots, y_d, 0, z_1, \dots, z_{i-2}, z_{i-1})$ is different from the number of time units to broadcast from $(y_i, \dots, y_d, 1, z_1, \dots, z_{i-2})$ to $(y_{i+1}, \dots, y_d, 1, z_1, \dots, z_{i-2}, z_{i-1})$. Both times units are either 1 or 2.

Let us have a look at the number of time units required to broadcast from $(y_1, \dots, y_d)$ to $(z_1, \dots, z_d)$ along the path P_k, $k \in \{0, 1\}$. The time t_k to broadcast the message via P_k is $t_k = t_{k_1} + t_{k_2} + \cdots + t_{k_{d+1}}$ with $t_{k_j} \in \{1, 2\}, 1 \le j \le d+1$. So we have

$$\sum_{k=0}^{1} t_k = (d+1)(1+2) = 3(d+1).$$

Since the path P_k are nodedisjoint besides of the first and the last node, the message will reach $(z_1, \dots, z_d)$ on one of these path at a time at most $3(d+1)/2$. $\square$

5.2.3   Lower Bounds for Bounded-Degree Graphs

In this subsection, the overall goal is to improve the lower bounds for broadcasting in the butterfly and the DeBruijn network. But in order to apply our proof techniques to other networks as well, we will concentrate on the general methods and arguments used and we will point out which properties of the graph we are using.

The first property which helps us improve the lower bound (at least for the DeBruijn graph) is that the graph we are considering has degree d. This argument was developed by Liestman and Peters [32] for graphs of degree 3 and 4 and further refined and sharpened for general d in [4] and [10]. As we are mainly interested in how the argument works, we only present the results for degree 3 and 4. The argument basically consists of finding an upper bound on the number of nodes which can be informed in t time steps.

Theorem 5.2.11 ([32])

    a) *Let G be a graph with n vertices and degree 3. Then*
$$b(G) \geq 1.4404 \log_2 n.$$

    b) *Let G be a graph with n vertices and degree 4. Then*
$$b(G) \geq 1.1374 \log_2 n.$$

Proof.

a) Let $A(t)$ denote the maximum number of nodes which can be newly informed in round t. Since G has degree 3, once a node has received the message it can only inform 2 additional neighbours in the next two rounds. Therefore, $A(t)$ is recursively defined as follows:

$$A(0) = 1, \ A(1) = 1, \ A(2) = 2, \ A(3) = 4,$$

$$A(t) = A(t-1) + A(t-2) \quad \text{for } t \geq 4.$$

For any broadcasting scheme running in time t,

$$\sum_{i=0}^{t} A(i) \geq n$$

must hold. A simple analysis shows that

$$A(i) \approx 1.6180^i \, ,$$

hence

$$\sum_{i=0}^{t} A(i) \approx 1.6180^t \geq n,$$

which yields $t \geq 1.4404 \log_2 n$.

b) Let $A(t)$ denote the maximum number of nodes which can be newly informed in round t. Since G has degree 4, once a node has received the message it can only inform 3 additional neighbours in the next three rounds.

Therefore, $A(t)$ is recursively defined as follows:

$$A(0) = 1, \; A(1) = 1, \; A(2) = 2, \; A(3) = 4, \; A(4) = 8,$$

$$A(t) = A(t-1) + A(t-2) + A(t-3) \quad \text{for } t \geq 5.$$

For any broadcasting scheme running in time t,

$$\sum_{i=0}^{t} A(i) \geq n$$

must hold. A simple analysis shows that

$$A(i) \approx 1.8393^i ,$$

hence

$$\sum_{i=0}^{t} A(i) \approx 1.8393^t \geq n,$$

which yields $t \geq 1.1374 \log_2 n$.

$\square$

For the butterfly network BF_k, Theorem 5.2.11 yields a lower bound of $1.1374k$ which is worse than the diameter lower bound. But for the DeBruijn network DB_k, we can improve the lower bound by applying Theorem 5.2.11:

Corollary 5.2.12
$b(DB_k) \geq 1.1374k.$

The technique of Liestman and Peters was extended by E. Stöhr [40] who was the first to prove a non-trivial lower bound of $1.5621k$ on the broadcast time of the butterfly network BF_k. Her technique was again refined and extended in [28], where the lower bound was improved to the currently best one of $1.7417k$. In order to make things easier to understand, we prove a slightly weaker bound. The graph property which is needed for the improvement is the following:

There is a node from which *a lot* of vertices have a *large* distance (*large* $\approx$ diameter).

The intuitive idea is that it is inherently difficult to send the message from the originating node to nodes very far away and to spread the information at the same time. This argument can be viewed as a generalization of Observation 5.2.3. It basically consists of finding an upper bound on the number of nodes which can be informed in t time steps at distance i. Taking also the distance into account is the difference to the technique of Liestman and Peters. As we will see, this makes the calculations much more difficult.

Let us start by stating the mentioned graph property more exactly for the butterfly network:

Lemma 5.2.13 *Let* BF_m *be the butterfly network of dimension* m. *Let* $v_0 = \langle 0, 00...0 \rangle$. *Let* $\varepsilon > 0$ *be any positive constant. Then there exist* $2^m - o(2^m)$ *nodes which are at distance at least* $\lfloor 3m/2 - \varepsilon m \rfloor$ *from* v_0.

Proof. Let

$$L = \{ \langle \lfloor m/2 \rfloor, \delta \rangle \mid \delta \neq \alpha 0^k \beta \text{ for some } k \geq \varepsilon m/2, \alpha 0^k \beta \in \{0,1\}^m \}$$

be the subset of the level-$\lfloor m/2 \rfloor$ vertices of BF_m. Then $|L| \geq 2^m - m 2^{m - \varepsilon m/2}$. It is not very difficult to show that the distance between any vertex v from L and v_0 is at least $\lfloor 3m/2 - \varepsilon m \rfloor$.   □

Now, we are able to show the improved lower bound:

Theorem 5.2.14 ([28])
   $b(BF_m) > 1.7396m$ *for all sufficiently large* m.

Proof. To obtain a contradiction suppose that broadcasting can be completed on BF_m in time $3m/2 + tm$, $0 \leq t < 1/2$.

As in the proof of Theorem 5.2.9, we can assume that the message originates at vertex $v_0 = \langle 0, 00...0 \rangle$, and the originator learns the message at time 0.

Let $A(i, t)$ denote the maximum number of nodes which can be reached in round t on a path of length i. Since BF_m has maximum degree 4, once a node has received the message it can only inform 3 additional neighbours in the next three rounds. Therefore, $A(i,t)$ is recursively defined as follows:

$A(0,0) = 1,$

$A(1,1) = 1,$

$A(1,2) = 1, \quad A(2,2) = 1,$

$A(1,3) = 1, \quad A(2,3) = 2, \quad A(3,3) = 1,$

$A(1,4) = 1, \quad A(2,4) = 3, \quad A(3,4) = 3, \quad A(4,4) = 1,$

$A(i,t) = A(i-1, t-1) + A(i-1, t-2) + A(i-1, t-3) \quad$ for $t \geq 5.$

It can easily be shown by induction (cf. [6]) that

$$A(n, n+l) \leq 2 \cdot \sum_{\substack{p+2q=l, \\ 0 \leq p, q \leq n}} \binom{n}{p+q} \cdot \binom{p+q}{q}.$$

Let $\varepsilon > 0$ be any positive constant. From Lemma 5.2.13, we know that for any broadcasting scheme

$$\sum_{n=3m/2-\varepsilon m}^{3m/2+tm} \sum_{l=0}^{3m/2+tm-n} A(n, n+l) \geq 2^m - o(2^m).$$

For ε tending towards 0, we have

$$2^m - o(2^m) \leq \sum_{n=3m/2}^{3m/2+tm} \sum_{l=0}^{3m/2+tm-n} A(n, n+l)$$

$$\leq \sum_{n=3m/2}^{3m/2+tm} \sum_{l=0}^{3m/2+tm-n} 2 \cdot \sum_{\substack{p+2q=l, \\ 0 \leq p, q \leq n}} \binom{n}{p+q} \cdot \binom{p+q}{q}$$

$$\leq 2 \cdot \sum_{n=3m/2}^{3m/2+tm} \sum_{0 \leq p+2q \leq 3m/2+tm-n} \binom{n}{p+q} \cdot \binom{p+q}{q}$$

$$\leq cm^3 \cdot \max_{\substack{3m/2 \leq n \leq 3m/2+tm, \\ 0 \leq p+2q \leq 3m/2+tm-n}} \binom{n}{p+q} \cdot \binom{p+q}{q}$$

for some constant c. It can easily be verified that the above maximum is obtained for $n = 3m/2$, $p + 2q = tm$ when $t < 1/2$. Therefore,

$$\max_{\substack{3m/2 \leq n \leq 3m/2 + tm, \\ 0 \leq p + 2q \leq 3m/2 + tm - n}} \binom{n}{p+q} \cdot \binom{p+q}{q} = \max_{0 \leq i \leq tm/2} \binom{3m/2}{tm-i} \cdot \binom{tm-i}{i}$$

The latter term is maximized for $i = i_0 m$ where

$$i_0 = \frac{1}{4} + \frac{t}{2} - \sqrt{\left(\frac{1}{4} + \frac{t}{2}\right)^2 - \frac{t^2}{3}}$$

For large m, an approximate expression for the factorial is given by Stirling's formula $m! \approx m^m e^{-m} \sqrt{2\pi m}$. Using Stirling's formula we obtain

$$\binom{3m/2}{tm - i_0 m} \cdot \binom{tm - i_0 m}{i_0 m}$$
$$\approx \frac{(3/2)^{3m/2}}{(3/2 - t + i_0)^{3m/2 - tm + i_0 m}(i_0)^{i_0 m}(t - 2i_0)^{tm - 2i_0 m}}.$$

Thus,

$$cm^3 \cdot \frac{(3/2)^{3m/2}}{(3/2 - t + i_0)^{3m/2 - tm + i_0 m}(i_0)^{i_0 m}(t - 2i_0)^{tm - 2i_0 m}} \geq 2^m - o(2^m).$$

Taking the m-th root on both sides, we have for large m

$$\frac{(3/2)^{3/2}}{(3/2 - t + i_0)^{3/2 - t + i_0}(i_0)^{i_0}(t - 2i_0)^{t - 2i_0}} \geq 2.$$

The latter inequality is not true for $t \leq 0.2396$. This contradiction establishes the theorem.   □

For the DeBruijn network, the proofs are completely analogous. First, we state the required graph property more exactly:

Lemma 5.2.15 *Let DB_m be the deBruijn network of dimension m. Let $v_0 = 00...0$. Let $\varepsilon > 0$ be any positive constant. Then there exist $2^m - o(2^m)$ nodes which are at distance at least $\lfloor m - \varepsilon m \rfloor$ from v_0.*

Proof. Let

$$L = \{ v \mid v \neq \alpha 0^k \beta \text{ for some } k \geq \varepsilon m, \, \alpha 0^k \beta \in \{0,1\}^m \}$$

be the subset of vertices. Then $|L| \geq 2^m - m 2^{m-\varepsilon m}$. Let $v \in L$. As the longest sequence α of consecutive 0's in v has at most length $\lfloor \varepsilon m \rfloor$, the bit string $v_0 = 00...0$ has to be rotated at least $\lfloor m - \varepsilon m \rfloor$ times to change the 1's left and right of α. Therefore, the distance between any vertex v from L and v_0 is at least $\lfloor m - \varepsilon m \rfloor$. $\qquad\square$

Now, we are able to state the improved lower bound:

Theorem 5.2.16 ([28])
$b(DB_m) > 1.3042m$ *for all sufficiently large m.*

Proof. The proof is similar to that of Theorem 5.2.14. We suppose that broadcasting can be completed on DB_m in time $m + tm$, $0 \leq t < 1/3$. The node $v_0 = 00...0$ is taken as the originator of the message. The recursion formula for $A(i, t)$ is exactly the same as in the proof of Theorem 5.2.14. This time, the condition obtained from Lemma 5.2.15 is

$$\sum_{n=m}^{m+tm} \sum_{l=0}^{m+tm-n} A(n, n+l) \geq 2^m - o(2^m).$$

Similar estimations as before show that this cannot be true for $t \leq 0.3042$. $\qquad\square$

As we already mentioned, the lower bounds in [28] are slightly better than the ones we proved, namely

$b(BF_k) \geq 1.7417k$ for all sufficiently large k,

$b(DB_k) \geq 1.3171k$ for all sufficiently large k.

To derive these bounds, a third property of the graph is used, namely:

Each edge is contained in a cycle of length at most 4.

This is true for the butterfly and the DeBruijn network. Using this additional property, the recursion for the number of informed nodes $A(i,t)$ is changed, and similar estimations as in the proofs above lead to the desired results. But the complete analysis is quite tedious and the improvement in the result is not very significant. Therefore, we omit this part here. We will rather give an overview on the broadcast time for small butterfly networks BF_k as displayed in Table 5.2.1.

k	lower bound	upper bound	no. processors
2	3	3	8
3	5	5	24
4	7	7	64
5	8	9	160
6	10	11	384
7	11	13	896
8	13	15	2048
9	15	17	4608
10	16	19	10240
11	18	21	22528
12	19	23	49152
13	21	25	106496
14	23	27	229376
15	24	29	491520
16	26	31	1048576
17	27	33	2228224
18	29	35	4718592

Table 5.2.1: Broadcast time for small butterfly networks BF_k

The upper bound is the $2k - 1$ algorithm from [28], the lower bound comes from the exact evaluation of the $A(i,t)$'s in the proof of Theorem 5.2.14 and the exact computation of the distances in the butterfly network. The overall picture is that the upper and lower bound are very close together for small dimensions k. For $k \leq 4$, i.e. up to 64 nodes, they even coincide.

As mentioned before, the effort to close the gap between the upper and the lower bound for BF_k and DB_k is still going on. The upper bound for BF_k has been improved to $2k - \frac{1}{2}\log\log k + c$, for some constant c and all sufficiently large k, in the meantime [41]. The upper bound for DB_k is still $\frac{3}{2}(k+1)$ [7]. It seems that it should be possible to improve the lower bounds. We think that there must be some other properties of the graphs which make broadcasting very difficult. One such property seems to be the many cycles of length k, but the analysis is not clear.

Finally, with some more extensions, it is possible to apply our lower bound technique to networks of higher node degree, e.g. general butterfly networks as considered in [2], or general deBruijn and Kautz networks as investigated in [7] and [26]. So far, we have been able to come up with the general analysis, which is somewhat more complicated. But we still have to apply it to the networks of interest.

5.2.4 Overview and Outlook

As a summary of this section, Table 5.2.2 contains an overview of the best currently known time bounds for broadcasting in common interconnection networks and the according references in this paper and in the literature.

graph	no. nodes	diameter	lower bound	upper bound
K_n	n	1	$\lceil \log_2 n \rceil$	$\lceil \log_2 n \rceil$
			Lemma 5.2.2	Lemma 5.2.2
H_k	2^k	k	k	k
			Lemma 5.2.2	Lemma 5.2.2
CCC_k	$k \cdot 2^k$	$\lfloor 5k/2 \rfloor - 1$	$\lceil 5k/2 \rceil - 2$	$\lceil 5k/2 \rceil - 2$
			Theo.5.2.7, [32]	Theo.5.2.7, [32]
SE_k	2^k	$2k - 1$	$2k - 1$	$2k$
			Theo.5.2.8, [24]	Theo.5.2.8, [24]
BF_k	$k \cdot 2^k$	$\lfloor 3k/2 \rfloor$	$1.7417k$	$2k - 1$
			Theo.5.2.14, [28]	Theo.5.2.9, [28]
DB_k	2^k	k	$1.3171k$	$\frac{3}{2}(k+1)$
			Theo.5.2.16, [28]	Theo.5.2.10, [7]

Table 5.2.2: Broadcast times for common networks

While improving the bounds for these networks is still of great interest, the search for new interconnection structures with better broadcasting capabilities is also going on. E.g. in [4], new families of graphs are presented achieving broadcasting in time $1.8750 \log_2 n$ and $1.4167 \log_2 n$ for degree 3 and 4, respectively, further closing the gap towards the lower bounds of $1.4404 \log_2 n$ and $1.1374 \log_2 n$ from Theorem 5.2.11. For general fixed degree, other constructions of networks with efficient broadcasting schemes are given in [10]. Broadcasting (and also gossiping) in the lately proposed star graph and pancake graph (as special csaes of Cayley graphs, see [1, 2]) has been investigated in [35, 3, 21] where efficient broadcast schemes are presented. The star graph and pancake graph have become very popular because they have a very regular interconnection pattern, sublogarithmic degree and diameter.

5.3   GOSSIPING

5.3.1   Introduction

This section is devoted to the gossip problem. Unlike the previous section we have to consider both one–way and two–way mode here. But, we shall mostly deal with the one–way mode because the two-way mode is so powerful in some cases that one can get optimal gossip algorithms in the two–way mode in a straightforward way (cf. gossiping examples with graphs G for which $r(G) = 2 \cdot minb(G)$ or $r(G) = r_2(G)$ or $r_2(G) = b(G)$, etc.).

It is impossible to give the proofs of all important and interesting results for gossiping here. So, we have chosen only some of them which represent some nice, principle proof ideas with possibly broader applications than only deriving the results presented.

Following the idea to start with simple results and to finish with harder results and open problems, we have structured this chapter as follows.

In Subsection 5.3.2 we give some optimal algorithms for weak–connected graphs (paths, trees, rings, etc.). Here we also find several graphs G with the property $r(G) = 2 \cdot minb(G)$ or $r(G) = r_2(G)$. Subsection 5.3.3 is devoted to optimal gossiping in complete graphs. The last subsection is concerned with gossiping in the most common interconnection networks. Some effective algorithms are presented which nicely use the optimal algorithms for gossiping in the ring as a subroutine. This subsection also includes the most interesting open problems formulated for gossiping.

Before starting Subsection 5.3.2 we give an interesting observation and some useful definitions and notations.

Observation 5.3.1 *Let $G = (V, E)$ be a graph with n nodes. Then*

$$r(G) \geq r_2(G) \geq \begin{cases} \lceil \log_2 n \rceil & \text{for even } n, \\ \lceil \log_2 n \rceil + 1 & \text{for odd } n. \end{cases}$$

Proof. $r(G) \geq r_2(G) \geq b(G)$ is clear. According to Observation 5.2.1, $b(G) \geq \lceil \log_2 n \rceil$ holds for any n.

Let n be odd. To see that $r_2(G) \geq \lceil \log_2 n \rceil + 1$, we first prove the following fact: "Let $E_1, E_2, \ldots, E_i$ be an arbitrary two–way communication algorithm of i rounds for a graph G. Then after the execution of this algorithm none of the nodes of G knows more than 2^i pieces of information originally distributed in G." Let us prove this fact by induction on i.

1. For $i = 0$ (no executed round) each node knows exactly $1 = 2^i$ piece of information originally residing in it.

2. Let the fact (hypothesis) hold for $k \leq m$. Let us show it also holds for $m + 1$. Let E_{m+1} be the $(m + 1)$-th round of a communication algorithm $E_1, E_2, \ldots, E_m, E_{m+1}, \ldots, E_s$. Following the induction hypothesis each node knows at most 2^m pieces of information after the m-th round. So, no node can get more than 2^m new pieces of information from another one in the $(m + 1)$-th round. Since $2^m + 2^m = 2^{m+1}$, each node knows at most 2^{m+1} pieces of information after the $(m + 1)$-th round.

Using the hypothesis above, we see that after $i = \lceil \log_2 n \rceil - 1$ rounds, each node of G knows at most $2^i = 2^{\lceil \log_2 n \rceil - 1}$ pieces of information. As the number n of nodes is odd, there is a node v_0 which cannot take part in any communication in round $\lceil \log_2 n \rceil$. Hence, v_0 has at most $2^{\lceil \log_2 n \rceil - 1} < n$ pieces of information after $\lceil \log_2 n \rceil$ rounds. $\square$

Now, the question appears whether there exist graphs of n nodes with the property $r_2(G) = \lceil \log_2 n \rceil$. These graphs are called **minimal gossip graphs**.

The answer is "yes". To show this we have to find a graph and a communication strategy which enables us to double the knowledge of each node in each round. We show that any graph of 2^m nodes having the hypercube H_m as a spanning subgraph has this property. In Section 5.3.3 we will give more examples of graphs having this property.

Lemma 5.3.2 *For any positive integer m:*

$$r_2(H_m) = m.$$

Proof. Obviously, $E(H_m)$ has exactly $m \cdot 2^{m-1}$ edges which can be distributed into m equal–sized disjoint sets $E_1, \ldots, E_m$, where $E_i = \{((\alpha_1, \alpha_2, \ldots, \alpha_{i-1}, 0, \alpha_{i+1}, \ldots, \alpha_m), (\alpha_1, \alpha_2, \alpha_{i-1}, 1, \alpha_{i+1}, \ldots, \alpha_m)) \mid \alpha_j \in \{0,1\}$ for $j = 1, 2, \ldots, i - 1, i + 1, \ldots, m\}$ for $i = 1, \ldots, m\}$. The sequence $E_1, \ldots, E_m$ (or any other permutation of this sequence) is a two–way gossip algorithm for H_m.

To see this it is sufficient to realize that by removing the edges in E_1 from H_m one gets two isolated $(m - 1)$–dimensional hypercubes H_{m-1}^0 and H_{m-1}^1, and that $(V(H_m), E_1)$ is a bipartite graph of degree 1. So, after the first round E_1, the nodes in H_{m-1}^k (for $k = 0, 1$) together know the whole cumulative message of H_m, and to complete the task it suffices to gossip in the $(m-1)$–dimensional Hypercubes H_{m-1}^0 and H_{m-1}^1. □

We note that there is no constant–degree interconnection network G of n nodes having the property $r_2(G) = \lceil \log_2 n \rceil$ because the doubling of the information is only possible if each node is active in each round via another edge. So, the degree $\lceil \log_2 n \rceil$ is necessary for each $r_2(G) = \log_2 n$.

We conclude this subsection with some definitions which enable us to see gossip algorithms from another point of view than as a sequence of communication rounds.

Definition 5.3.3 *Let $G = (V, E)$ be a graph, and let $X = x_1, \ldots, x_m$ be a simple path (i.e. $x_i \neq x_j$ for $i \neq j$) in G. Let $A = E_1, \ldots, E_s$ $[R_1, \ldots, R_s]$ be a communication algorithm in two–way [one–way] mode. Let $T = t_1, \ldots, t_{m-1}$ be an increasing sequence of positive integers such that $(x_i, x_{i+1}) \in E_{t_i}$ $[(x_i \to x_{i+1}) \in R_{t_i}]$ for $i = 1, \ldots, m - 1$. We say that $X[t_1, \ldots, t_{m-1}]$ is a time–path of A because it provides the information flow x_1 to x_m in A. If $t_{i+1} - t_i - 1 = k_i \geq 0$ for some $i \in \{1, \ldots, m - 2\}$ then we say that $X[t_1, \ldots, t_{m-1}]$*

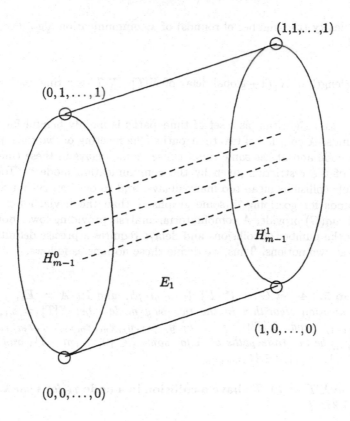

Figure 5.22

has a k_i-**delay** at the node x_{i+1}. The **global delay** of $X[t_1, \ldots, t_{m-1}]$ is $d(X[t_1, \ldots, t_{m-1}]) = t_1 - 1 + \sum_{i=1}^{m-2} k_i$. The **global time** of $X[t_1, \ldots, t_{m-1}]$ is $m - 1 + d(X[t_1, \ldots, t_{m-1}])$.

Obviously, the necessary and sufficient condition for a communication algorithm to be a gossip algorithm in a graph G is the existence of time–paths among all nodes in G.

So, one can view the gossip algorithm for a graph G as a set of time–paths between any ordered pair of nodes.

The complexity (the number of rounds) of a communication algorithm can be measured as

$$\max\{\text{length of } X[T] + \text{global delay of } X[T] \mid X[T] \text{ is a time-path of } A\}.$$

To see a gossip algorithm as a set of time–paths is mainly helpful for proving lower bounds. A collision of two time–paths (the meeting of two time–paths at the same node and at the same time) causes some delays in these time–paths (because of the restriction given by the communication modes). Too many unavoidable collisions mean too many delays, and so one can get much better lower bounds for gossiping in some graphs G than the trivial lower bounds $d(G)$ and $rad(G)$ provide. A combinatorial analysis providing lower bounds by analyzing the number of collisions and delays requires a precise definition and use of these two notions. Thus, we define these notions as follows.

Definition 5.3.4 *Let $G = (V, E)$ be a graph, and let $A = E_1, \ldots, E_s$ be a communication algorithm in the two–way mode. Let $X[T] = x_1, \ldots, x_m$ $[t_1, \ldots, t_{m-1}]$ and $Y[T'] = y_1, y_2, \cdots, y_j, x_m, x_{m-1}, \ldots, x_{m-i+1}, x_{m-i}$ $[t'_1, \ldots, t'_{j+i}]$ be two time–paths of A for some $i \in \{0, \ldots, m-1\}$, and $y_l \neq x_r$ for all $l \in \{1, \ldots, j\}, r \in \{1, \ldots, m\}$.*

*We say that $X[T]$ and $Y[T']$ **have a collision in a node** x_k for some $k = m-d$ (see Fig 5.23) if*

(i) $t_{k-2} = t'_{j+d-1}$ *or* $\max\{t_{k-2}, t'_{j+d-1}\} < \min\{t_{k-1}, t'_{j+d}\}$

> *{ The condition (i) predetermines the conflict in the node x_k, because (i) ensures that messages flowing via time-paths $X[T]$ and $Y[T']$ have reached the nodes x_{k-1} and x_{k+1} resp. before any of them has reached the node x_k.*
> *}*

(ii) $t_{k-1} < t'_{j+d}$ *or* $t'_{j+d} < t_{k-1}$

> *{ These two conditions describe all possible solutions of the conflict predetermined by (i).}*

If $t_{k-1} < t'_{j+d}$ we say that **the collision causes a 1–delay on the time–path** $Y[T']$. *If $t'_{j+d} < t_{k-1}$ we say that* **the collision causes a 1–delay on the time–path** $X[T]$.

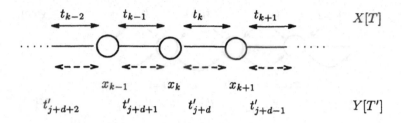

Figure 5.23

Next we define the collisions for one–way communication algorithms. It can easily be observed that a collision in this communication mode causes at least one 2–delay on one of the two time–paths in the collision or a 1–delay on each of these two paths.

Definition 5.3.5 *Let $G = (V, E)$ be a graph, and let $A = E_1, \ldots, E_s$ be a communication algorithm in the one–way mode. Let $X[T] = x_1, \ldots, x_m$ $[t_1, \ldots, t_{m-1}]$, and $Y[T'] = y_1 y_2 \cdots y_j x_m x_{m-1} \cdots x_{m-i}$ $[t'_1, \ldots, t'_{j+i}]$ be two time–paths of A for some $i \in \{1, \ldots, m-1\}$ and $y_l \neq x_r$ for all $l \in \{1, \ldots, j\}, r \in \{1, \ldots, m\}$. We say that $X[T]$ and $Y[T']$ have a collision in a node x_k for some $k = m - d$ (see Fig. 5.24) if*

(i) $t_k > t'_{j+d+1} > t_{k-1} > t'_{j+d}$, or

(ii) $t'_{j+d+1} > t_k > t'_{j+d} > t_{k-1}$, or

(iii) $t_k > t'_{j+d+1} > t'_{j+d} > t_{k-1}$, or

(iv) $t'_{j+d+1} > t_k > t_{k-1} > t'_{j+d}$

If one of the cases (i) or (ii) happens, then we say that **the collision causes a 1–delay on** $X[T]$ **and a 1–delay on** $Y[T']$. *If (iii) happens, then we say that* **the collision causes a 2–delay on** $X[T]$. *If (iv) happens, then we say that* **the collision causes a 2–delay on** $Y[T']$.

We note that our Definitions 5.3.4 and 5.3.5 are not the only possibilities how to formalize the notions collisions and delays caused by collisions. If some combinatorial analysis requires a definition of collisions which covers more conflicts

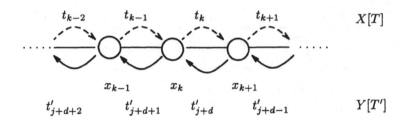

Figure 5.24

appearing in communication algorithms, then one can define this in several distinct ways. One of such broader definitions can be found in [24] in the proof of the precise lower bound for gossiping in rings. Since we prefer to give some basic proof ideas here rather than to present too detailed, technical proofs, the Definitions 5.3.4 and 5.3.5 will be sufficient for the purpose followed here.

5.3.2   Gossiping in Graphs with Weak Connectivity

We shall investigate the gossip problem for weak–connected graphs in this subsection. The reason to do this at first is that such graphs are relatively simple (there are not many disjoint paths in weak–connected graphs) and so the gossip problem in such structures is easier to analyse. We start with the simple case–paths.

Theorem 5.3.6

1. $r_2(P_n) = n - 1$ for any even integer $n \geq 2$,

2. $r_2(P_n) = n$ for any odd integer $n \geq 3$,

3. $r(P_n) = n$ for any even integer $n \geq 2$ and

4. $r(P_n) = n + 1$ for any odd integer $n \geq 3$.

Proof.

(1) $r_2(P_n) \geq b(P_n) \geq d(P_n) = n - 1$. Let $V(P_n) = \{x_1, \ldots, x_n\}$. The following gossip algorithm $\{(x_1, x_2), (x_{n-1}, x_n)\}$, $\{(x_2, x_3)(x_{n-2}, x_{n-1})\}$, $\ldots$, $\{(x_{n/2-1}, \quad x_{n/2}), \quad (x_{n/2+1}, x_{n/2+2})\}$, $\{(x_{n/2}, x_{n/2+1})\}$, $\{(x_{n/2-1}, x_{n/2}), (x_{n/2+1}, x_{n/2+2})\}$, $\ldots$, $\{(x_1, x_2), (x_{n-1}, x_n)\}$ works in $n-1$ rounds. Another gossip algorithm working in $n - 1$ rounds is $A = E_1, E_2, \ldots, E_{n-1}$, where $E_i = \{(x_1, x_2), (x_3, x_4), \ldots, (x_{n-1}, x_n)\}$ for odd i, and $E_j = E(P_n) - E_1$ for all even j.

(2) $r_2(P_n) \leq n$ follows from the following gossip algorithm $A' = E_1', E_2', \ldots, E_{n-1}', E_n'$, where $E_i' = \{(x_1, x_2), (x_3, x_4), \ldots, (x_{n-2}, x_{n-1})\}$ for i odd, and $E_j' = E(P_n) - E_1'$ for all even j. To prove $r_2(P_n) \geq n$, let us consider the paths $\vec{X} = x_1, \ldots, x_n$ and $\overleftarrow{X} = x_n, \ldots, x_1$. Obviously each gossip algorithm for P_n must contain two time–paths $\vec{X}[T]$ and $\overleftarrow{X}[T']$ for some $T = t_1, \ldots, t_{n-1}$ and $T' = t_1', \ldots, t_{n-1}'$. Obviously, it is sufficient to prove that at least one of the two time–paths $\vec{X}[T]$ and $\overleftarrow{X}[T]$ has the global delay of at least 1 (note that the length of $\vec{X}$ is $n - 1$). Let us assume that there is no positive delay on the time–paths $x_1, x_2, \ldots, x_{\lceil n/2 \rceil - 1}[t_1, \ldots, t_{\lceil n/2 \rceil - 2}]$ and $x_n, x_{n-1}, \ldots, x_{\lceil n/2 \rceil + 1}[t_1', \ldots, t_{\lceil n/2 \rceil - 2}']$ (in the opposite case the proof is already finished), i. e. that $t_i = t_i' = i$ for $i = 1, \ldots, \lceil n/2 \rceil - 2$. Following Definition 5.3.4 we see that there must be a collision between $\vec{X}[T]$ and $\overleftarrow{X}[T']$ in the node $x_{\lceil n/2 \rceil}$ (see Fig.5.23). Thus at least one of $\vec{X}[T]$ and $\overleftarrow{X}[T']$ has a positive global delay, i. e. the maximum of the global time of $\vec{X}[T]$ and the global time of $\overleftarrow{X}[T']$ is at least n.

(3)(4) The gossip algorithms showing $r(P_n) \leq 2 \cdot \lceil n/2 \rceil$ for any $n \geq 2$ can be derived by combing Example 5.1.12 and Lemma 5.1.13. Any one–way gossip algorithm for P_n must contain the time–paths $X[T] = x_1, \ldots, x_n[t_1, \ldots, t_{n-1}]$ and $X^R[T] = x_n, \ldots, x_1[t_1', \ldots, t_{n-1}']$. Because these time–paths are going in the opposite direction on the same path $x_1, \ldots, x_n$ there must be a collision between $X[T]$ and $X^R[T']$ at some node x_i. Obviously this implies $r(P_n) \geq n$ for any $n \geq 2$.

Now, let us consider the case n is odd. Again, a collision occurs at some node x_i. If $x_i \neq x_{\lceil n/2 \rceil}$, the one of the time–paths already has a delay before the collision, and we are done because any collision in the one–way mode requires either a 2–delay for some time–path or a 1–delay for each of the two paths in collision. Let us assume that $x_i = x_{\lceil n/2 \rceil}$ and $t_j = t_j' = j$ for $j = 1, 2, \ldots, \lceil n/2 \rceil - 2$ (see Fig. 5.24). Then, the collision in $x_{\lceil n/2 \rceil}$ has one of the 4 types (i), (ii), (iii), (iv) of Definition 5.3.5. Obviously (iii) and (iv) causes a 2–delay on one ot the time–paths which completes the proof. Now, let us consider the case (i) (the case (ii) is analogous), where $\lceil n/2 \rceil - 1 = t_{\lceil n/2 \rceil - 1} < t'_{\lceil n/2 \rceil - 1} < t_{\lceil n/2 \rceil} < t'_{\lceil n/2 \rceil}$. Obviously, $t'_{\lceil n/2 \rceil - 1} - t'_{\lceil n/2 \rceil - 2} > 1$ and $t'_{\lceil n/2 \rceil} - t'_{\lceil n/2 \rceil - 1} > 1$. Thus, this collision causes at least two positive delays on the path $X^R[T']$.

$\square$

The next graphs for which we show optimal one–way gossiping are complete k-ary trees. To prove this we present the following lemma.

Lemma 5.3.7 ([5]) $r(T) = 2 \, minb(T)$ for any tree T.

Proof. From Lemma 5.1.13, we have $r(T) \leq 2 \cdot minb(T)$ for any tree T. To show $r(T) \geq 2 \, minb(T)$, let $A = E_1, \ldots, E_s$ be any one–way gossip algorithm for T, and let t_A be the first round after which at least one node of T knows the cumulative message. Obviously $t_A \geq minb(T)$. Let $V(t_A)$ be all nodes having the cumulative message after t_A rounds. We show that $|V(t_A)| = 1$ by contradiction.

Let $u, v \in V(t_A), u \neq v$. Because T is a tree there exists exactly one path $u, y_1, \ldots, y_k, v$ (k may be 0) between u and v (see Fig. 5.25). Let T_v be the subtree rooted at v excluding the edge (v, y_k) and the subtree rooted at y_k. Let T' be the subtree rooted at y_k excluding the edge (y_k, v) and the tree T_v. Now, let $t \leq t_A$ be the last round in which (y_k, v) was used for communication in the first t_A rounds of A. Now, we distinguish two possibilities depending on the direction of the communication.

1. Let $(y_k \to v) \in E_t$. Then y_k must already know the cumulative message of T' after the $(t-1)$-th round (if not then v cannot know the cumulative message of T after t_A rounds because all pieces of information originally residing in T' can flow to v only via y_k). Since $(y_k \to v)$ is the last use of the edge (y_k, v) in $E_1, E_2, \ldots, E_{t_A}$ (i. e. no information from T_v will flow to u later in the rounds $E_t, \ldots, E_{t_A}$), any piece of information originally distributed in T_v can flow to u only via y_k, and u must know the cumulative message of T_v after t_A rounds of A, y_k must already know the cumulative message of T_v before the t-th round (after the last round containing $(v \to y_k)$). Thus, we get that y_k knows the whole cumulative message of T (union of the cumulative messages of T' and T_v) already before the t-th round. But, this is a contradiction to the assumption that no node knows the cumulative message after $t_A - 1$ rounds.

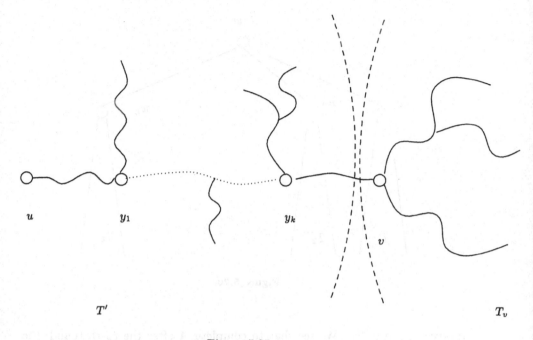

Figure 5.25

2. Let $(y_k \leftarrow v) \in E_t$. In the same way as in the first case it can be shown that v must already learn the cumulative message of T before the t_A-th round, which is again a contradiction.

Now, we may assume $V(t_A) = \{w\}$ for a node w in T. Let us view w as the root of T with k sons $w_1, w_2, \ldots, w_k$ (see Fig. 5.26) for some positive integer k.

Let T_i denote the subtree rooted at w_i for $i = 1, \ldots, k$. Since w knows the cumulative message after t_A rounds, each w_i knows the cumulative message of T_i (note that each information flowing from a node in T_i to w must flow via w_i), and no node in T_i knows a piece of information which is unknown to w_i (note that each piece of information flowing to a node in T_i from a node outside of T_i must flow via w_i) for any $i \in \{1, \ldots, k\}$. On the other hand, none of the nodes $w_1, w_2, \ldots, w_k$ knows the cumulative message of T after t_A rounds. Thus, for each $i \in \{1, \ldots, k\}$, there is a piece of information $p(i)$ which is unknown

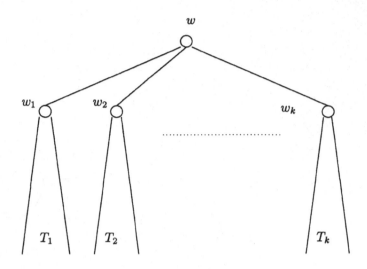

Figure 5.26

to every node in T_i. We see that to complete A after the t_A-th round, the remaining rounds $E_{t_A+1}, \ldots, E_s$ must contain the broadcast of $p(1)$ from w in T_1, the broadcast of $p(2)$ from w in T_2, etc. Obviously, this cannot be easier than broadcasting from w in T. Thus, $s = t_A + (s - t_A) \geq minb(T) + minb(T) \geq 2\, minb(T)$.   □

So, following Lemma 5.3.7 and Lemma 5.2.4, we get the following result.

Theorem 5.3.8

$$r\,(T_k^m) = 2\, minb\,(T_k^m) = 2 \cdot k \cdot m \quad \textit{for all integers } m \geq 1, k \geq 2. \quad \square$$

Now, the question appears whether there exist non–tree graphs G with $r(G) = 2\, minb(G)$. The answer is yes and the next technical lemma enables us to find such graphs. The proof of this lemma is a generalization of the idea used in the proof of Lemma 5.3.7.

Lemma 5.3.9 *Let G be a graph with a bridge (v, u) (i.e. an edge) whose removal (from G) divides G into two components G_1 and G_2. Then*

$$r(G) \geq minb(G) + 1 + \min\{minb(G_1), minb(G_2)\}.$$

Proof. Let T be the time unit in which at least one node of G has learned the whole cumulative message (i.e., all pieces of information distributed in G), and no node of G knows the cumulative message in the time unit $T - 1$ (i.e. after $T - 1$ rounds). Let $G = (V, E), G_1 = (V_1, E_1), G_2 = (V_2, E_2), u \in V_1, v \in V_2$, and let V_T be the set of all nodes that know the whole cumulative message after T rounds. We shall prove that either $V_T \subseteq V_1$ or $V_T \subseteq V_2$.

Let us prove this fact by contradiction. Let there exist two nodes $v_1 \in V_1 \cap V_T$ and $v_2 \in V_2 \cap V_T$. Since $v_1(v_2)$ knows all pieces of information distributed in $G_2(G_1)$ after T rounds, and the whole information exchange between G_1 and G_2 flows through the edge (v, u), the whole cumulative message has flown through the edge (v, u) in the first T rounds. So, the nodes v and u belong to V_T. But this is impossible because when the last information exchange between u and v was from $u(v)$ to $v(u)$ in a round $T' \leq T$ then $u(v)$ has learned the cumulative message already before this information exchange (i.e., before the round T) [see Fig. 5.27].

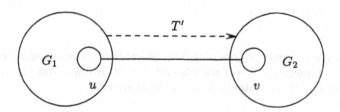

Figure 5.27

So, we have proved that either $V_T \subseteq V_1$ or $V_T \subseteq V_2$. W.l.o.g. let us assume that $V_T \subseteq V_1$. Since the nodes in $V_T \subseteq V_1$ know all pieces of information distributed in G_2 we have that the node $v \in V_2$ must also know all pieces of information distributed in G_2. Since $v \notin V_T$, v does not learn at least one piece of information distributed in G_1 in the first T rounds. So, we need at least $1 + b_v(G_2)$ rounds to distribute this piece of information in G_2. Clearly, in the case $V_T \subseteq V_2$ we need at least $1 + b_u(G_1)$ rounds to finish the gossiping after T rounds. Since $T \geq minb(G)$ we obtain the claimed inequality. $\square$

Now, let us show that Lemma 5.3.9 provides optimal lower bounds for gossiping on some infinite class of graphs.

Let us consider two cycles R_1 and R_2, each with n nodes, n even, connected by one edge (u, v) [see Fig. 5.28].

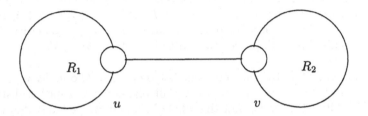

Figure 5.28

Using Example 5.1.12, $minb(R_1) = minb(R_2) = \frac{n}{2}$, and $minb(G) = \frac{n}{2} + 1$. So, applying Lemma 5.3.9 we obtain

$$r(G) \geq n + 2.$$

An optimal algorithm for gossiping first concentrates the cumulative information in u in $minb(G) = \frac{n}{2} + 1$ rounds and then it disseminates the cumulative message from u to all nodes in G in $minb(G)$ rounds.

So, for two connected cycles of the same size we have proved $r(G) = 2\,minb(G)$, i.e. we have found graphs different from trees with the property that gossiping is exactly two times harder than broadcasting. We note that Lemma 5.3.9 provides optimal lower bounds for $r(G)$ of several further graphs (see, for example, some trees, cycles connected by one simple path, etc.). Many of them also have the property $r(G) = 2\,minb(G)$. (To be more precise, all of them for which $minb(G) = b_u(G_1) + 1 = b_v(G_2) + 1$)

Now, let us present a version of Lemma 5.3.9 providing lower bounds for the two-way communication mode.

Lemma 5.3.10   *Let G be a graph with a bridge (v, u) whose removal divides G into two components G_1 and G_2. Then*

$$r_2(G) \geq minb(G) + \min\{minb(G_1), minb(G_2)\}.$$

Sketch of the proof: Similarly as in the proof of Lemma 5.3.9 it can be proved that either $V_T \subseteq V_i$ for some $i \in \{1, 2\}$ or that $V_T = \{v, u\}$. $V_T = \{v, u\}$ exactly holds in the case when v and u make an information exchange in the T-th round.

Clearly, if $V_T \subseteq V_i$ for some $i \in \{1, 2\}$ then $r_2(G) \geq T + 1 + minb(G_j) \geq minb(G) + minb(G_j) + 1$, $j \in \{1, 2\} \setminus i$. If $V_T = \{v, u\}$ then $r_2(G) \geq T + \max\{minb(G_1), minb(G_2)\}$. So $r_2(G) \geq minb(G) + \min\{1 + minb(G_1), 1 + minb(G_2), \max\{minb(G_1), minb_2(G_2)\}\}$. $\square$

Considering the two connected cycles R' from Fig. 5.28, Lemma 5.3.10 implies an optimal lower bound $r_2(R') \geq n + 1$. So R' is an interesting example because the two-way mode decreases the complexity of gossiping only by 1 (note that both upper bounds of Lemma 5.1.13, $r(R') = 2 \cdot minb(R')$ and $r_2(R') = 2 \cdot minb(R') - 1$ are satisfied).

Next, we shall establish the exact values for $r(C_n)$ and $r_2(C_n)$. This is of importance because we shall show in Subsection 3.4 some algorithm for gossiping in some prominent interconnection networks which effectivity depends strongly on some subroutine arranging the gossiping in cycles. While to find an optimal two–way gossip algorithm for the cycle C_n is a simple task, the one–way version of this task is already hard. Since the lower bound proof for $r_2(C_n)$ in [24] takes more than 10 pages, we do not present this detailed combinatorial analysis here. But we illustrate the proof idea based on the analysis of collisions by proving a weaker lower bound in a shorter way. First, we start with the result for $r_2(C_n)$.

Theorem 5.3.11 ([19])

$$r_2(C_k) = k/2 \text{ for even } k \geq 4, \text{ and } r_2(C_k) = \lceil k/2 \rceil + 1 \text{ for odd } k \geq 3.$$

Proof.   Let us give the proof only for k even. The case for k odd is left as an exercise for the reader.

Obviously, $\text{rad}(C_k) = k/2$ and so $r_2(C_k) \geq k/2$. Let $V(C_k) = \{x_1, \ldots, x_k\}$. The gossip algorithm for C_k is $A = E_1, E_2, \ldots, E_{k/2}$, where $E_i = \{(x_1, x_2), (x_3, x_4), \ldots, (x_{k-1}, x_k)\}$ for all odd i, and $E_j = \{(x_2, x_3), (x_4, x_5), \ldots, (x_{k-2}, x_{k-1}), (x_k, x_1)\}$ for all even j. To see that A is a gossip algorithm it is sufficient to realize that after i rounds each node knows exactly $2i$ pieces of information.   □

Next, we present the optimal one-way gossip algorithm in cycles of even length established in [24]. Note that slightly weaker lower and upper bounds on $r(C_k)$ have been established in [12].

Theorem 5.3.12 ([24])

$$r(C_n) = n/2 + \lceil \sqrt{2n} \rceil - 1 \text{ for each even } n > 3, \text{ and}$$

$$\lceil n/2 \rceil + \lceil \sqrt{2n - 1/2} \rceil - 1 \leq r(C_n) \text{ each odd } n \geq 2\sqrt{\lceil n/2 \rceil} - 1$$

Proof. Let us first prove the upper bounds.

1. To explain the idea we first give the algorithm for $n = 2l^2$, l even. Then we extend the algorithm for any positive integer n. Let us divide the cycle C_n into l disjoint paths of lengths $2l$, the i-th path starting with v_i and ending with u_i, as depicted in Fig. 5.29. Let $v_i'(u_i')$ be a node between v_i and u_i with the distance $l - 1$ from $v_i(u_i)$. Now, the algorithm works in two phases.

   1st Phase   For each $i \in \{1, \ldots, l\}$:
   there is a time-path of length $n/2$ from v_i to $v_{(i+l/2-1) \bmod l+1}$ going through u_i, and
   there is a time-path of length $n/2 - 1$ from u_{i-1} to $v_{(i+l/2-1) \bmod l+1}$ going through v_{i-1}.
   (Clearly, the time-paths starting in v_i's go in opposite direction as the time-paths starting in u_i's.)

   Note that after the 1st phase all nodes v_i already know the cumulative message because for each v_i there are two time-paths: one from $v_{(i+l/2-1) \bmod l+1}$ to v_i and the second one from $u_{(i+l/2-1-1) \bmod l+1}$ to v_i.

   2nd Phase   For each $i \in \{1, \ldots l\}$:
   v_i sends the cumulative message to u_{i-1}
   v_i sends the cumulative message to v_i',
   u_i sends the cumulative message to u_i',

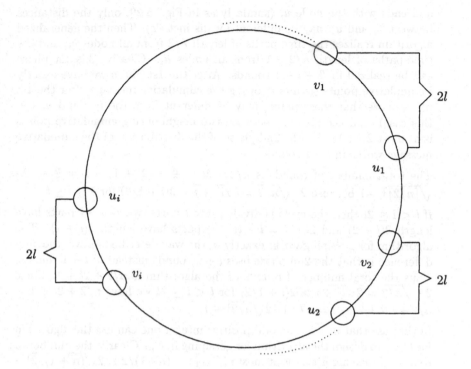

Figure 5.29

Now, let us add the number of rounds. Each time-path starting in a v_i in the 1st phase has length $n/2$ and it has exactly $(n/2l) - 1 = l - 1$ collisions. When the collision of two time-paths is solved in such a way that the collision causes a 1-delay for each time-path, then the 1st phase uses $n/2 + l - 1$ rounds. Since the distance between v_i and u_i is $2l - 1$, the 2nd phase uses l rounds. One can simply see that $n/2 + 2l - 1 = n/2 + \lceil \sqrt{2n} \rceil - 1$.

Now, let us give an algorithm for even n. For each even, positive integer $n > 3$ there is a positive integer l such that

$$2l^2 \leq n < 2(l+1)^2 = 2l^2 + 4l + 2$$

Thus $n = 2l^2 + 2i$ for some $i \in \{0, 1, \ldots, 2l\}$. If $1 \leq i \leq l$ then we divide the cycle into such l parts $P_1, \ldots, P_l$ that i parts have length $2(l+1)$ and $l - i$ parts have length $2l$. For each i the part P_i starts with the node v_i

and ends with the node u_i (similarly as in Fig. 5.29, only the distances between v_i and u_i may be different for distinct i's). Then the generalized algorithm realizes the time-paths of length $n/2$ from all nodes v_i, and the time-paths of length $n/2 - 1$ from all nodes u_i. Clearly, this 1st phase can be realized in $\frac{n}{2} + l - 1$ rounds. After the 1st phase we have exactly l cumulative points (points knowing the cumulative message after the 1st phase, note that these points may be different from the v_i's and u_i's in this case), and the distance between two neighbouring cumulative points is at most $2(l+1)$. So, the 2nd phase of the distribution of the cumulative message works in $l + 1$ rounds.

The total number of rounds is $n/2 + 2l = \frac{n}{2} + (2l + 1) - 1 = \frac{n}{2} + \lceil 2 \cdot \sqrt{\lceil n/2 \rceil} \rceil - 1$ because $2 \cdot \sqrt{n/2} = 2\sqrt{l^2 + i} < 2(l + 1/2)$ for $1 \le i \le l$.

If $l < i \le 2l$ then the cycle is divided into l parts, where $i - l$ parts have length $2(l + 2)$ and $2l - i = l - (i - l)$ parts have length $2(l + 1)$. The algorithm for gossiping works exactly in the way described above, the only difference is that the 2nd phase uses $l + 2$ rounds instead of $l + 1$ rounds. Thus the total number of rounds of the algorithm is $n/2 + 2l + 1$. Since $2 \cdot \sqrt{n/2} = 2\sqrt{l^2 + i} > 2(l + 1/2)$ for $l < i \le 2l$ we have $n/2 + 2l + 1 = n/2 + (2l + 2) - 1 = n/2 + \lceil 2\sqrt{n/2} \rceil - 1$.

In the case that $n > 1$ is an odd positive integer one can use the algorithm for C_{n+1} to design the algorithm for gossiping in C_n. Clearly, the number of rounds of such algorithm is at most $r(C_{n+1}) = (n+1)/2 + \lceil 2\sqrt{(n + 1)/2} \rceil - 1 = \lceil n/2 \rceil + \lceil 2\sqrt{\lceil n/2 \rceil} \rceil - 1$.   □

2. Let us now deal with the lower bound. Because the complete proof of the optimal lower bound [24] requires too many specific considerations taking a lot of space, we shall present here only the following weaker lower bound:

$$r(C_n) \ge n/2 + \sqrt{2n}/4 - O(1).$$

The proof of this lower bound will be sufficient for learning more about the combinatorial lower bound proof technique based on the investigation of collisions. If somebody wants to know the complete power of this techniques for rings the paper [24] should be consulted.

Since the optimal algorithm for gossiping in C_n in the two–way mode uses at least $\lceil n/2 \rceil$ rounds (Theorem 5.3.11) we may assume that $r(C_n) = \lceil n/2 \rceil + f(n)$ for some function f from positive to nonnegative integers. Next we shall show that $f(n) \ge \sqrt{2n}/4 - O(1)$.

Let A be an arbitrary optimal algorithm for gossiping in C_n in one–way mode. Let A work in $t(A) = \lceil n/2 \rceil + f(n)$ rounds. From the upper bound on $r(C_n)$ we know $f(n) \leq \lceil \sqrt{2n} \rceil$. Thus, each time-path of A has the global time at most $\lceil n/2 \rceil + f(n)$, i.e., there is no time–path in A for paths longer than $\lceil n/2 \rceil + f(n)$. It implies that any two nodes x and y lying at distance $\lfloor n/2 \rfloor - f(n) - 1$ must have two time-paths in A, one leading from x to y and another one going from y to x, both realized on the shortests path between x and y (see Fig. 5.30).

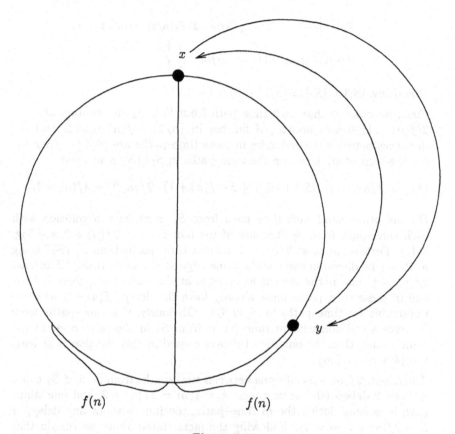

Figure 5.30

Let us now consider the sets of time–paths

$$S_1 = \left\{ x_1, x_2, \ldots, x_{\lfloor n/2 \rfloor - f(n)}[T_1]; x_2, x_3, \ldots, x_{\lfloor n/2 \rfloor - f(n) + 1}[T_2]; \right.$$
$$\left. \cdots; \quad x_{\lfloor n/2 \rfloor - 3f(n) - 2}, x_{\lfloor n/2 \rfloor - 3f(n) - 1}, \cdots, \right.$$

$$x_{2\lfloor n/2\rfloor - 4f(n) - 3}\left[T_{\lfloor n/2\rfloor - 3f(n)}\right]\Big\}$$

and

$$S_2 = \Big\{ x_{\lfloor n/2\rfloor - f(n)}, x_{\lfloor n/2\rfloor - f(n) - 1}, \dots, x_1[T_1']; x_{\lfloor n/2\rfloor - f(n) + 1}, \dots,$$

$$x_2[T_2']; \dots; \quad x_{2\lfloor n/2\rfloor - 4f(n) - 3}, x_{2\lfloor n/2\rfloor - 4f(n) - 2}, \dots,$$

$$x_{\lfloor n/2\rfloor - 3f(n) - 2}\left[T_{\lfloor n/2\rfloor - 3f(n) - 2}'\right]\Big\}$$

Obviously, $|S_1| = |S_2| = \lfloor n/2\rfloor - 3f(n) - 2$.

First, we observe that each time–path from $S_1 \cup S_2$ can contain at most $2(f(n) + 1)$ delays because A finishes in $\lceil n/2\rceil + f(n)$ rounds and the distances between the endnodes in these time–paths are $\lfloor n/2\rfloor - f(n) - 1$. So, the sum of all delays on the time–paths in $S_1 \cup S_2$ is at most

(1) $\quad 2\left(f(n) + 1\right) \cdot \left(|S_1| + |S_2|\right) \le 2 \cdot \left(f(n) + 1\right) \cdot 2\left(\lfloor n/2\rfloor - 3f(n) - 2\right).$

On the other hand each time–path from S_1 must have a collision with each time–path from S_2 (because of the fact $i - j = 2f(n) + 2$, see Fig. 5.31). Further, at most $2f(n) + 2$ distinct time–paths from S_1 (S_2) going in the same direction can use the same edge at the same time. (If at least $2f(n) + 4$ time–paths use the same edge at the same time, then at least one of these time–paths must already have the delay $2f(n) + 3$ which is impossible for time–paths in $S_1 \cup S_2$). Obviously, if k time–paths from S_1 have a collision with m time–paths from S_2 in the same node in the same round, then the number of delays caused in this collision is at least $\min\{k + m, 2k, 2m\}$.

Thus, each $2f(n) + 2$ collisions between time–paths from S_1 and S_2 cause at least 2 delays (the worst cases: $k = 1, m = 2f(n) + 2$, and one time–path is waiting during the m time–paths continue without any delay, or $k = 2f(n) + 2, m = 1$). Following the facts stated above we obtain that the sum of all delays on the time–paths from $S_1 \cup S_2$ is at least

(2) $\qquad\qquad 2 \cdot \dfrac{(|S_1| \cdot |S_2|)}{2f(n) + 2} = \dfrac{(\lfloor n/2\rfloor - 3f(n) - 2)^2}{f(n) + 1}.$

Comparing (1) and (2) we get

$$\frac{(\lfloor n/2\rfloor - 3f(n) - 2)^2}{f(n) + 1} \le 2\left(f(n) + 1\right) \cdot 2\left(\lfloor n/2\rfloor - 3f(n) - 2\right)$$

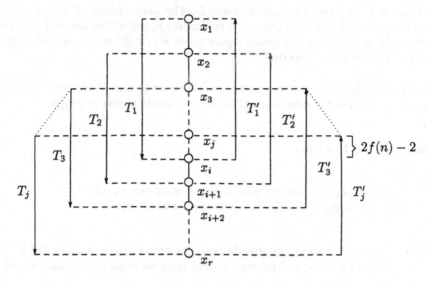

Figure 5.31 $j = \lfloor n/2 \rfloor - 3f(n) - 2, i = \lfloor n/2 \rfloor - f(n), r = j + i - 1$

$$2\left(f(n) + 1\right)^2 \geq \frac{\left(\lfloor n/2 \rfloor - 3f(n) - 2\right)^2}{2\left(\lfloor n/2 \rfloor - 3f(n) - 2\right)} = \frac{1}{2} \cdot \left(\lfloor n/2 \rfloor - 3f(n) - 2\right)$$

which gives the lower bound $f(n) \geq \frac{\sqrt{2}}{4} \cdot \sqrt{n} - O(1)$.

$\square$

Concluding this section we note that the optimal gossip algorithms for cycles can be used to obtain further optimal gossip algorithms for some classes of weak–connected graphs. Some results of this kind are presented in [23, 24]. Another application leading to effective gossiping in some interconnection networks will be shown in Subsection 5.3.4.

5.3.3 Gossiping in Complete Graphs

The aim of this subsection is to present optimal one-way and two-way gossip algorithms in K_n. While the gossip problem in K_n is relatively simply solvable in two-way mode, the design of the optimal one-way gossip algorithm in K_n

requires a little more elaborated method. The presentation of this method, counting precisely the necessary and sufficient growth of the amount of information disseminated in any gossip algorithm in K_n, is the main methodological contribution of this subsection.

We consider first the two-way communication mode and show that for every natural number n the complete graph K_n is a minimal gossip graph.

Theorem 5.3.13 ([30])

$r_2(K_n) = \lceil \log_2 n \rceil$ *for every positive, even integer* n, *and*
$r_2(K_n) = \lceil \log_2 n \rceil + 1$ *for every positive, odd integer* n.

Proof. The lower bound has already been presented in Observation 5.3.1. We start with the upper bound for even n, and then we reduce the case for odd n to this case.

First, we have to show that $r_2(K_n) \leq \lceil \log_2 n \rceil$ holds for even n. Let $n = 2m$. We partition the set of processors into two sets Q, R of size m. Let us denote the processors by $Q[i], R[i], 0 \leq i \leq m - 1$. The following algorithm doubles the information at each node in every step.

Algorithm 2-WAY-GOSSIP-K_n

    <u>for</u> all $i \in \{0, \ldots, m - 1\}$ <u>do</u> in parallel
        exchange information between $Q[i]$ and $R[i]$;
    <u>for</u> $t = 1$ to $\lceil \log_2 m \rceil$ <u>do</u>
    <u>for</u> all $i \in \{0, \ldots, m - 1\}$ <u>do</u> in parallel
        exchange information between $Q[i]$ and $R[(i + 2^{t-1}) \bmod m]$;

Let $q[i], r[i], 0 \leq i \leq m-1$, denote the pieces of information stored by processors $Q[i], R[i]$ before starting the algorithm. Set $\alpha[i] = \{q[i], r[i]\}, 0 \leq i \leq m - 1$.

After the execution of the first instruction, processors $Q[i]$ and $R[i]$ both store $\alpha[i]$.   It   is   easy   to   verify by induction on t that after round $t, 1 \leq t \leq \lceil \log_2 m \rceil$, processors $Q[i]$ and $R[(i + 2^{t-1}) \bmod m]$ both store the set of pieces of information

$$\bigcup_{0 \leq j \leq 2^t - 1} \alpha[(i + j) \bmod m]$$

Therefore, after $1 + \lceil \log_2 m \rceil = \lceil \log_2 n \rceil$ rounds, all nodes have received the complete information.

Now, let $n = 2m + 1$. Number the nodes of K_n from 1 to n. The following algorithm performs gossiping in K_n:

1. Send the information of the node $i + m$ to the node i for all $2 \leq i \leq m + 1$. {After this step, the cumulative message is distributed in the nodes $1, 2, \ldots, m + 1$.}

2. If $m + 1$ is even, gossip in $1, 2, \ldots, m + 1$. If not, gossip in $1, 2, \ldots, m + 2$. {After this step, each of the nodes $1, 2, \ldots, m + 1$ knows the cumulative message.}

3. Send the information of the node i to the node $i + m$ for all $2 \leq i \leq m + 1$. {After this step, each of the nodes knows the cumulative message.}

If $m + 1$ is even, the above algorithm takes

$$r_2(K_{m+1}) + 2 = \lceil \log_2(m + 1) \rceil + 2 = \lceil \log_2 \left(\tfrac{n+1}{2} \right) \rceil + 2$$
$$= \lceil \log_2(n + 1) \rceil + 1 = \lceil \log_2 n \rceil + 1$$

rounds. If $m + 1$ is odd, then $n + 1$ is not a power of two, and hence the algorithm takes

$$r_2(K_{m+2}) + 2 = \lceil \log_2(m + 2) \rceil + 2 = \lceil \log_2 \left(\tfrac{n+3}{2} \right) \rceil + 2$$
$$= \lceil \log_2(n + 3) \rceil + 1 = \lceil \log_2 n \rceil + 1$$

rounds.   □

The algorithm described in Theorem 5.3.13 does not use all the edges of the complete graph. In fact, since the algorithm uses only $\lceil \log_2 n \rceil$ rounds (we consider here only the case where n is even), for every node at most $\lceil \log_2 n \rceil$

of its edges are used. Thus the algorithm defines a graph of degree at most $\lceil \log_2 n \rceil$. We call this graph Gossip graph and denote it by Gos_n.

Gos_n is defined for even $n, n = 2m$, and has n nodes which are denoted by $Q[i]$ and $R[i]$, $0 \leq i \leq m - 1$. The edges connect $Q[i]$ and $R[i]$ for every i, $0 \leq i \leq m - 1$, and furthermore for every i, $0 \leq i \leq m - 1$, and for every t, $1 \leq t \leq \lceil \log_2 m \rceil$, there are edges connecting $Q[i]$ with $R[(i + 2^{t-1}) \mod m]$. The graph Gos_{12} is shown in Fig. 5.32.

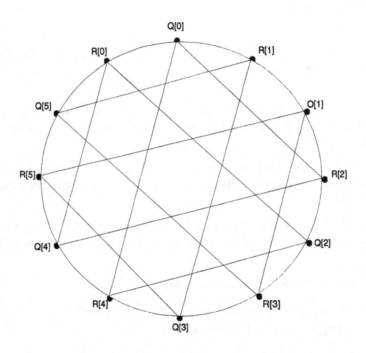

Figure 5.32   The graph Gos_{12}

Because of the construction of Gos_n the following result follows from Theorem 5.3.13.

Corollary 5.3.14 *For every even positive integer n:*

$$r_2(Gos_n) = \lceil \log_2 n \rceil.$$

It is not difficult to see that if n is a power of 2, i.e. $n = 2^k$ for some k, then the graph Gos_n is isomorphic to H_k, the hypercube of dimension k. Thus, a special consequence of Corollary 5.3.14 is that the hypercube is a minimal gossip graph (which was already shown in Lemma 5.3.2). We had already seen that the hypercube is very well-suited for information dissemination in Lemma 5.2.2, where we showed that the hypercube is a minimal broadcast graph. Of course, for even n every minimal gossip graph is also a minimal broadcast graph and Corollary 5.3.14 can be viewed as a generalization of Lemma 5.2.2.

Finding an optimal gossip strategy in the one-way mode is more sophisticated. The number of rounds in this case is determined by the growth of the Fibonacci numbers which are defined by

$$F(1) = F(2) = 1,$$
$$F(n + 1) = F(n) + F(n - 1) \text{ for } n \geq 2.$$

Let $b = \frac{1}{2}(1 + \sqrt{5})$. Using $b^2 = b + 1$, it is easily provable by induction that $b^{i-2} \leq F(i) \leq b^{i-1}$ holds for all $i \geq 2$ [29].

We will consider the gossip problem in the one-way mode only for an even number of nodes. Results for odd number of nodes can be found in [14].

Theorem 5.3.15 ([15]) *For every positive, even integer n, and for every positive integer k with $F(k) \geq n/2$:*

$$r(K_n) \leq k + 1.$$

Proof. The algorithm is somewhat similar to the algorithm presented in the proof of Theorem 5.3.13. Again the set of processors is partitioned into two equal-sized subsets Q and R. In each round either all processors from Q send their information to all processors from R or vice versa, i.e. in each round either all processors from Q are writing and all processors from R are reading, or all processors from Q are reading and all processors from R are writing. Let $n = 2m$, and let us denote the processors by $Q[i], R[i], 0 \leq i \leq m - 1$.

Algorithm 1-WAY-GOSSIP-K_n

$t := 0;$
for all $i \in \{0, \ldots, m - 1\}$ do in parallel

         $R[i]$ sends to $Q[i]$;
    for all $i \in \{0, \ldots, m-1\}$ do in parallel
         $Q[i]$ sends to $R[i]$;
    while $F(2t+1) < m$ do
    begin
       $t := t+1$;
       for all $i \in \{0, \ldots, m-1\}$ do in parallel
          $R[(i + F(2t-1))$ mod $m]$ sends to $Q[i]$;
       if $F(2t) < 2m$ then
       for all $i \in \{0, \ldots, m-1\}$ do in parallel
          $Q[(i + F(2t))$ mod $m]$ sends to $R[i]$
   end;

When this algorithm stops [since $F(2t+1) \leq m$ or $F(2t) \geq m$ holds], then it has performed $2t$ rounds or $2(t-1)+1$ rounds, respectively, within the while loop. Therefore, the algorithm performs $(k-1)+2$ rounds, where k is the smallest integer such that $F(k) \geq m$.

In order to prove the correctness of the algorithm, let again $q[i], r[i], 0 \leq i \leq m-1$, denote the pieces of information stored by processors $Q[i], R[i]$ before starting the algorithm. Set $\alpha[i] = \{q[i], r[i]\}, 0 \leq i \leq m-1$.

After the execution of the first two instructions processors $Q[i]$ and $R[i]$ both store $\alpha[i]$. It is not difficult to verify by induction on t that after t runs through the while loop of the above algorithms

$$Q[i] \text{ stores } \bigcup_{0 \leq j \leq F(2t+1)-1} \alpha[(i+j) \text{ mod } m],$$

$$R[i] \text{ stores } \bigcup_{0 \leq j \leq F(2t+2)-1} \alpha[(i+j) \text{ mod } m].$$

If k is an odd number, $k = 2t+1$, then after t runs of the while loop all processors store the whole information. If k is an even number, $k = 2t+2$, then also the first instruction in run $t+1$ has to be executed before all processors store the whole information.   ☐

While this upper bound was published already in 1979, it took a long time before it was proved to be optimal. Perhaps this was due to the fact that people did not believe in the optimality of the algorithm. The partition into two sets, the first two rounds (which are just the simulation of one round in the 2-way mode) and the static distinction between senders and receivers seem to leave a lot of freedom for further improvements. In 1988/89 four groups, [12, 14, 33, 42], independently found a way to prove the optimality. The methods they use are very similar. We describe here the approach from [14].

For this purpose, a new problem is introduced which is called Network Counting Problem (NCP).

The information stored by each of the n processors will be an integer. At the beginning all the integers will be equal to one. The processors are communicating in the one-way mode, i.e. in each round either a processor sends its integer or it receives an integer. If it receives an integer, then it adds this integer to its own integer. Again we are interested in the number of rounds needed until all processors store an integer which is greater or equal to n.

It is clear that any algorithm for solving the gossip problem also solves NCP and that a lower bound for NCP is also a lower bound for the gossip problem. There exists a straightforward algorithm for solving NCP. The set of processors is partitioned into groups, each of two processors. Within such a group the processors alternately send their information to each other, i.e. after t rounds one of them stores $F(t+1)$ and the other one $F(t+2)$.

Therefore the above algorithm needs $k-1$ rounds, where k is the smallest integer such that $n \leq F(k)$.

We shall prove now the lower bound and we shall do this by proving a lower bound for the Network Counting Problem (NCP). We will see that one round of the NCP can be described by some associated matrix and we shall use methods from matrix theory to prove the lower bound. We shall give here the necessary definitions, for a more elaborated description the reader is referred to [43].

Let $||..||$ be any vector norm over $\mathbb{R}^n$, i.e. $||x|| = 0 \Leftrightarrow x = 0^n, ||\alpha \cdot x|| = |\alpha| \cdot ||x||, ||x+y|| \leq ||x|| + ||y||$ for $\alpha \in \mathbb{R}, x, y \in \mathbb{R}^n$.

The matrix norm associated to a vector norm $||..||$ is defined by
$$||A|| = sup_{x \neq 0} \frac{||Ax||}{||x||}.$$

This matrix norm fulfills the following properties:

$$||A|| = 0 \Leftrightarrow A = 0$$
$$||A + B|| \leq ||A|| + ||B||$$
$$||\alpha A|| = \alpha \cdot ||A||$$
$$||A \cdot B|| \leq ||A|| \cdot ||B||$$
$$||A \cdot x|| \leq ||A|| \cdot ||x||$$

for all $A, B \in \mathbb{R}^{n^2}, x \in \mathbb{R}^n, \alpha \in \mathbb{R}, \alpha \geq 0$.

It turns out that for proving our lower bounds the Euclidean vector norm, defined by $||x|| = \sqrt{\Sigma_{i=1}^n |x_i|^2}, x = (x_1, .., x_n)$, is appropriate. It is well-known that the spectral norm is associated to this vector norm as a matrix norm. $||A|| = $ spectral norm$(A) = \sqrt{|\lambda_{max}(A^T \cdot A)|}$ where A^T is the transposed matrix of A and λ_{max} denotes the eigenvalue of maximal absolute value.

Let us consider now the NCP and let us consider one round in an algorithm solving the NCP. Let $u, v \in \mathbb{N}^n$ be the vectors of numbers stored by the n processors before and after that round. Associate to this round an $n \times n$ matrix A with entries $a_{i,j} \in \{0, 1\}$ by

(i) $a_{ii} = 1 \qquad \forall i = 1, .., n$
(ii) $a_{ij} = 1$, $i \neq j \Leftrightarrow$ processor j sends its number to processor i.

Then $Au = v$ holds and since we are working in the one-way mode A fulfills the following properties:

(i) $a_{ii} = 1 \qquad \forall i = 1, .., n$
(ii) $a_{ij} = 1$, $i \neq j \Rightarrow a_{iv} = a_{vj} = 0 \qquad \forall v \neq i, j$, $a_{vi} = 0 \quad \forall v \neq$
i , $a_{jv} = 0 \quad \forall v \neq j$

We will denote this class of matrices by $P(1, 1, n)$ and we have to determine the largest spectral norm $||A||$ among all matrices $A \in P(1, 1, n)$.

Note that every matrix $A \in P(1, 1, n)$ can be transformed by using coordinate transformations into a matrix with "blocks" of the form $B = \begin{pmatrix} 1 & 1 \\ 0 & 1 \end{pmatrix}$ along the main diagonal, i.e.

$$
TAT^{-1} = \begin{pmatrix}
B & & & & & & 0 \\
& B & & & & & \\
& & \ddots & & & & \\
& & & B & & & \\
& & & & 1 & & \\
& & & & & \ddots & \\
0 & & & & & & 1
\end{pmatrix}
$$

The spectral norms of A, TAT^{-1} and B coincide and the spectral norm of B is easily computable.

$$
B^T \cdot B = \begin{pmatrix} 1 & 0 \\ 1 & 1 \end{pmatrix} \begin{pmatrix} 1 & 1 \\ 0 & 1 \end{pmatrix} = \begin{pmatrix} 1 & 1 \\ 1 & 2 \end{pmatrix}
$$
$$
\Rightarrow (2 - \lambda)(1 - \lambda) - 1 = 0
$$
$$
\Rightarrow \lambda^2 - 3\lambda + 1 = 0
$$
$$
\Rightarrow \lambda_{max}(B^T B) = \frac{3}{2} + \sqrt{\frac{5}{4}}
$$

$$
\Rightarrow \|A\| = \sqrt{\lambda_{max}(A^T A)} = \frac{1}{2}(1 + \sqrt{5})
$$

Theorem 5.3.16 ([12, 14, 33, 42]) *Let n be an even positive integer. Every algorithm for solving NCP in the telegraph communication mode needs at least $2 + \lceil \log_b \frac{n}{2} \rceil$ rounds, where $b = \frac{1}{2}(1 + \sqrt{5})$.*

Proof. Let there exist a solution of r rounds, let $A_i, 1 \leq i \leq r$, be the matrix associated to round i of the algorithm. Let $\alpha_i, 1 \leq i \leq n$, be the number of pieces of information gathered by processor i during the first $r - 2$ rounds of the algorithm, i.e.

$$\alpha := \begin{pmatrix} \alpha_1 \\ \cdot \\ \cdot \\ \cdot \\ \alpha_n \end{pmatrix} = A_{r-2} \cdot \ldots \cdot A_2 \cdot A_1 \cdot \begin{pmatrix} 1 \\ \cdot \\ \cdot \\ \cdot \\ 1 \end{pmatrix}$$

$$\Rightarrow \|\alpha\| \leq \left(\prod_{i=1}^{r-2} \|A_i\|\right) \cdot \|(1, ..., 1)\| \leq b^{r-2} \cdot \sqrt{n}$$

Let us denote by $inf(i,t)$ the number of pieces of information gathered by processor i in the first t rounds of the algorithm. Since this algorithm needs r rounds, $inf(i,r) \geq n$ for all $i = 1,.., n$.

In the last round at most $\frac{n}{2}$ processors can gather more information, i.e. $inf(i,r-1) \geq n$ for at least $\frac{n}{2}$ processors i. There may exist already some indices i such that $\alpha_i = inf(i,r-2) \geq n$ holds. But, if $\alpha_i < n$ and $inf(i,r-1) \geq n$, then there exists some processor j with $\alpha_i + \alpha_j \geq n$ sending its information in round $r - 1$ to processor i. We distinguish three cases:

(1) $\alpha_i \geq n$,

(2) $\alpha_i < n$ and $\alpha_j \geq n$,

(3) $\alpha_i < n, \alpha_j < n$ and $\alpha_i + \alpha_j \geq n$.

Let c_k be the number of indices for which $(k), 1 \leq k \leq 3$, holds. Then $c_1 \geq c_2$ and $c_1 + c_2 + c_3 \geq \frac{n}{2}$, and therefore $2c_1 + c_3 \geq \frac{n}{2}$ holds.

Furthermore we use that for arbitrary numbers $\beta, \gamma \in \mathbb{R}$ with $\beta + \gamma = n$ the expression $\beta^2 + \gamma^2$ has the minimal value for $\beta = \gamma = \frac{n}{2}$. Putting all this together we get the following estimation:

$$\|\alpha\| = \sqrt{\sum_{i=1}^n \alpha_i^2} \geq \sqrt{c_1 n^2 + c_3 \cdot 2 \cdot \frac{n^2}{4}} \geq n \cdot \sqrt{\frac{1}{2}(2c_1 + c_3)} \geq \frac{n}{2}\sqrt{n}$$

We have shown an upper and a lower bound for $\|\alpha\|$, i.e. $\frac{n}{2} \cdot \sqrt{n} \leq \|\alpha\| \leq b^{r-2} \cdot \sqrt{n}$.
This implies $r \geq 2 + \lceil \log_b \lceil \frac{n}{2} \rceil \rceil$. □

The upper bound from Theorem 5.3.15 and the lower bound from Theorem 5.3.16 are very close together. The following lemma shows that their difference is at most 1 and that they are equal for infinitely many n.

Lemma 5.3.17 *Let $n = 2m$ be some even integer, and let $t_1 := 1 + k$ [where k is the smallest integer such that $m \leq F(k)$] be the upper bound from Theorem 5.3.15 and $t_2 := 2 + \lceil \log_b m \rceil$ [where $b = \frac{1}{2}(1 + \sqrt{5})$] be the lower bound from Theorem 5.3.16.*
Then $t_1 = t_2$ holds for infinitely many m and $t_1 \leq t_2 + 1$ holds for all m.

Proof. b fulfills $b^2 = b + 1$ and we have already mentioned that this implies $b^{i-2} \leq F(i) \leq b^{i-1}$ for all $i \geq 2$ [29].

Consider $n \in \mathbb{N}$ such that $n = 2 \cdot F(k)$ for some k. Then $t_1 = k + 1$ and $t_2 = 2 + \lceil \log_b F(k) \rceil = 2 + k - 1 = k + 1$. Therefore $t_1 = t_2$ holds for such n.

Let $n = 2m$ be an arbitrary, positive integer. If i is determined by $b^{i-1} < m \leq b^i$, then $t_2 = 2 + i$.
Let k be the smallest positive integer such that $F(k) \geq m$. Since $b^{k-2} \leq F(k) \leq b^{k-1}$, we obtain either $i = k - 1$ or $i = k - 2$. This implies $t_1 = k + 1 \leq i + 3$.
□

Now we know that the difference between the upper bound and the lower bound is at most one and this makes us more ambitious. We would like to know the exact value. The following Table 5.3.1 shows the upper bound for the gossip problem and the upper and lower bound for the network counting problem for numbers up to $n = 22$.

n	2	4	6	8	10	12	14	16	18	20	22
upper bound gossip	2	4	5	6	6	7	7	7	8	8	8
upper bound NCP	2	4	5	5	6	6	7	7	7	7	8
lower bound NCP	2	4	5	5	6	6	7	7	7	7	7

Table 5.3.1: Upper and lower bounds for Gossip and NCP

We have shown (the proof is omitted here) that solving the gossip problem for 8 processors needs 6 rounds and this shows that there exist integers n for which the network counting problem has smaller complexity than the gossip problem. For $n = 22$ the upper bound and the lower bound of the network counting problem also differ. It seems to be a solvable problem to determine the exact complexity of this problem.

5.3.4   Gossiping in Interconnection Networks

In this subsection we give a short survey on gossiping in the most familiar interconnection networks. In some cases we also present the best–known gossip algorithms in order to show some interesting ideas for the design of communication algorithms.

The hypercube. The hypercube is one of most popular parallel architectures, and so the investigation of communication problems for it is of great importance. We already have shown that $r_2(H_k) = k$, but to estimate $r(H_k)$ seems to be much more difficult. There are several one–way gossip algorithms working in $2k$ round (see, for instance [11]) (one of them can be directly obtained from the two–way gossip algorithm working in $2k$ rounds), and some researchers have also conjectured that $r(H_n) = 2n$ [5]. Surprisingly, Krumme [31] has found a one–way gossip algorithm for H_9 working in 17 rounds. The generalization of this algorithm for larger hypercubes has lead to a gossip algorithm working in $1.88k$ rounds [31]. We note that the highest known lower bound on $r(H_k)$ follows from Subsection 5.3.3; it leads to $r(H_k) \geq 1.44k$. This gap between $1.44k$ and $1.88k$ leaves enough space for further investigation. We note that we do not have any conjecture concerning the placement of $r(H_k)$ between $1.44k$ and $1.88k$, i.e., we do not know whether there is a greater chance to improve the lower bound than to improve the upper bound or vice versa.

Cube Connected Cycles (CCC) and Butterfly (BF). CCC_k and BF_k are important constant degree networks designed by some "transformations" of the hypercube with the aim to conserve the nice properties of the hypercube, and to decrease the degree of H_n.

Now, we present the best–known gossip algorithms for CCC_k and BF_k [23]. These algorithms are based on the combination of two ideas. One is the optimal gossip algorithm in cycles working in two phases (in the first one some nodes accumulate the whole cumulative message of the cycle, and in the second phase these nodes broadcast the cumulative message to the other nodes), and the second idea is so–called "set to set broadcasting" introduced in [37]. Let us explain set to set broadcasting. Let A and B be two sets of nodes. The set to set broadcasting from A to B is a communication process in which each node in B has learned all pieces of information distributed in A. In what follows we give algorithms for set to set broadcasting from the i-th level to the i-th level in BF_k, and for set to set broadcasting from the i-th level to the $((i-1) \bmod k)$-th level in CCC_k.

SET CCC_k

> for $j = 0$ to $k - 1$ do
>> for all $\alpha \in \{0,1\}^k$ do in parallel
>>> begin
>>>> exchange information between
>>>> $((i+j) \bmod k, \alpha)$ and $((i+j) \bmod k, \alpha((i+j) \bmod k))$ { * needs two rounds * };
>>>> if $j < k - 1$ then
>>>>> $((i+j) \bmod k, \alpha)$ sends to $((i+j+1) \bmod k, \alpha)$ { * needs 1 round * }
>>> end;

SET BF_k

> for $j = 0$ to $k - 1$ do
>> for all $\alpha \in \{0,1\}^k$ do in parallel
>>> begin
>>>> $((i+j) \bmod k, \alpha)$ sends to $((i+j+1) \bmod k, \alpha((i+j) \bmod k))$
>>>> $((i+j) \bmod k, \alpha)$ sends to $((i+j+1) \bmod k, \alpha)$
>>> end;

Now, gossiping in CCC_k and BF_k can be done as follows.

Algorithm GOSSIP-BF_k

1. Use in parallel for all α the 1st phase of the optimal algorithm for gossiping in C_k concentrating the cumulative message of C_k in $l = \lfloor \sqrt{\lceil k/2 \rceil} \rfloor$ "regularly distributed" nodes in C_k to concentrate the cumulative message of $C_\alpha(k)$ of BF_k in l nodes (v_i, α) for $1 \le i \le l$.

2. For all $i \in \{v_j | 1 \le j \le l\}$ do in parallel set to set broadcasting from the i-th level to the i-th level on BF_k.

3. Use in parallel for all α the 2nd phase of the optimal algorithm for gossiping in the cycle to broadcast the cumulative message of BF_k contained in the nodes $(v_i, \alpha), 1 \le i \le l$, to the other nodes in the cycle $C_\alpha(k)$.

Algorithm GOSSIP-CCC_k

$1°$ Use in parallel for all $\alpha \in \{0,1\}^{2^k}$ the 1st phase of the optimal algorithm for gossiping in C_k to concentrate the cumulative message of $C_\alpha(k)$ of CCC_k in $l = \lfloor \sqrt{\lceil k/2 \rceil} \rfloor$ nodes (v_i, α) for $1 \le i \le l$.

$2°$ For all $i \in \{v_j | 1 \le j \le l\}$ do in parallel set to set broadcasting from the i-th level of CCC_k to the $((i-1) \mathrm{mod}\, k)$-th level of CCC_k.

$3°$ Use in parallel for all α the 2nd phase of the optimal algorithm for gossiping in the cycle to broadcast the cumulative message of CCC_k contained in the nodes $((v_i - 1) \mathrm{mod}\, k, \alpha)$ of $CCC_k, 1 \le i \le l$, to the other nodes in the cycle $C_\alpha(k)$.

Analyzing the complexity of the above stated procedures we obtain:

Theorem 5.3.18 ([23]) *For every integer $k \ge 3$:*

$$r(CCC_k) \le r(C_k) + 3k - 1 \le \lceil 7k/2 \rceil + \lceil 2\sqrt{\lceil k/2 \rceil} \rceil - 2, \text{ and}$$
$$r(\mathrm{BF}_k) \le r(C_k) + 2k \le \lceil 5k/2 \rceil + \lceil 2\sqrt{\lceil k/2 \rceil} \rceil - 1.$$

Now, we show that this technique can be used also for two–way gossiping in CCC_k.

Theorem 5.3.19 ([23]) *Let $k \ge 3$ be an integer. Then*

$$r_2(CCC_k) \le k/2 + 2k = 5 \cdot \lceil k/2 \rceil \text{ for } k \text{ even, and}$$
$$r_2(CCC_k) \le \lceil k/2 \rceil + 2k + 2 = 5 \cdot \lceil k/2 \rceil \text{ for } k \text{ odd.}$$

Proof. To do gossiping in CCC_k the following algorithm working in three phases can be used.

1. Use the optimal algorithm for gossiping in C_k in two-way mode [FP80] to do gossiping in parallel on all cycles $C_\alpha(k)$ of CCC_k.

2. For all odd $i \leq k - 1$ do in parallel set to set broadcasting from the i-th level to the $((i - 1) \bmod k)$-th level on CCC_k.

3. For all odd $j \leq k - 1$ do in parallel:
   the j-th levels learns in parallel from the $(j - 1)$-th level [the $(k - 1)$-th level learns in parallel in one special round when k is odd].

The result of Theorem 5.3.19 follows directly from the fact $r_2(C_k) = k/2$ for k even and $r_2(C_k) = \lceil k/2 \rceil + 1$ for k odd proved in [FP80] and from the fact that the information exchange in the algorithm $SET\ CCC_k$ performed in the two-way mode runs in one round.   □

As CCC_k is a subgraph of BF_k [20], we have the following corollary for two–way gossiping in BF_k:

Corollary 5.3.20 ([23]) *Let $k \geq 3$ be an integer. Then*

$$r_2(BF_k) \leq k/2 + 2k = 5 \cdot \lceil k/2 \rceil \text{ for } k \text{ even, and}$$
$$r_2(BF_k) \leq \lceil k/2 \rceil + 2k + 2 = 5 \cdot \lceil k/2 \rceil \text{ for } k \text{ odd.}$$

Shuffle-Exchange (SE) and DeBruijn (DB). We do not know any non-trivial gossip algorithm for SE_k or DB_k in the one-way mode. The trivial algorithms based on the concatenation of the best known broadcast algorithms and accumulation algorithms yield $r(SE_k) \leq 4k - 2$, $r_2(SE_k) \leq 4k - 3$ (Theorem 5.2.8), and $r(DB_k) \leq 3k + 3$, $r_2(DB_k) \leq 3k + 2$ (Theorem 5.2.10). Since both SE_k and DB_k have many nice structural properties, there is hope for much better gossiping algorithms for these networks. So, this is also one of the most challenging problems for further research.

The grid. Multidimensional grids have relatively large diameter, and this makes it possible to perform one–way gossip optimally in the number of rounds equal to the diameter. This result is established for $n_1 \times n_2 \times \cdots \times n_k$ grids for any $k \geq 2$ and any $n_i \geq 9$ [12]. In [5] it is shown that for $n \times m$ grids, where $n \geq 6$ and m is even (odd), one–way gossip is possible in one round (two rounds) more than the diameter. For several small grids the problem to find optimal gossip algorithm is still left open [12].

5.3.5 Overview

As a summary of this section, Tables 5.3.2 and 5.3.3 contain overviews of the best currently known time bounds for gossiping in the one–way and two-way modes for common interconnection networks and the according references in this paper and in the literature.

In the tables, $even(n) = 1$ if n is even and 0 else, and $odd(n) = 1$ if n is odd and 0 else. Most of the lower bounds derive from the lower bounds for broadcasting.

graph	no. nodes	diameter	lower bound	upper bound
K_n	n	1	$\lceil \log_2 n \rceil + odd(n)$ [30]	$\lceil \log_2 n \rceil + odd(n)$ [30]
H_k	2^k	k	k Lemma 5.3.2	k Lemma 5.3.2
P_n	n	$n-1$	$n - even(n)$ Theo.5.3.6	$n - even(n)$ Theo.5.3.6
C_n	n	$\lfloor n/2 \rfloor$	$\lceil n/2 \rceil + odd(n)$ Theo.5.3.11, [19]	$\lceil n/2 \rceil + odd(n)$ Theo.5.3.11, [19]
CCC_k	$k \cdot 2^k$	$\lfloor 5k/2 \rfloor - 2$	$\lceil 5k/2 \rceil - 2$ Theo.5.2.7, [32]	$5 \cdot \lceil k/2 \rceil$ Theo.5.3.19, [23]
SE_k	2^k	$2k - 1$	$2k - 1$ Theo.5.2.8, [24]	$4k - 3$ Theo.5.2.8, [24]
BF_k	$k \cdot 2^k$	$\lfloor 3k/2 \rfloor$	$1.7417k$ Theo.5.2.14, [28]	$5 \cdot \lceil k/2 \rceil$ Cor.5.3.20, [23]
DB_k	2^k	k	$1.3171k$ Theo.5.2.16, [28]	$3k + 2$ Theo.5.2.10, [7]

Table 5.3.2: Gossip times for common networks in the two–way mode

graph	no. nodes	diameter	lower bound	upper bound
K_n	n	1	$1.44 \log_2 n$ [14]	$1.44 \log_2 n$ [15, 14]
H_k	2^k	k	$1.44k$ [14]	$1.88k$ [31]
P_n	n	$n-1$	$n + odd(n)$ Theo.5.3.6	$n + odd(n)$ Theo.5.3.6
C_n	n even	$\lfloor n/2 \rfloor$	$n/2 + \lceil \sqrt{2n} \rceil - 1$	$n/2 + \lceil \sqrt{2n} \rceil - 1$
	n odd	$\lfloor n/2 \rfloor$	$\lceil n/2 \rceil + \lceil \sqrt{2n - 1/2} \rceil - 1$ Theo.5.3.12, [24]	$\lceil n/2 \rceil + \lceil 2\sqrt{\lceil n/2 \rceil} \rceil - 1$ Theo.5.3.12, [24]
CCC_k	$k \cdot 2^k$	$\lfloor 5k/2 \rfloor - 2$	$\lceil 5k/2 \rceil - 2$ Theo.5.2.7, [32]	$\lceil 7k/2 \rceil + \lceil 2\sqrt{\lceil k/2 \rceil} \rceil - 2$ Theo.5.3.18, [23]
SE_k	2^k	$2k-1$	$2k-1$ Theo.5.2.8, [24]	$4k-2$ Theo.5.2.8, [24]
BF_k	$k \cdot 2^k$	$\lfloor 3k/2 \rfloor$	$1.7417k$ Theo.5.2.14, [28]	$\lceil 5k/2 \rceil + \lceil 2\sqrt{\lceil k/2 \rceil} \rceil - 1$ Theo.5.3.18, [23]
DB_k	2^k	k	$1.3171k$ Theo.5.2.16, [28]	$3k+3$ Theo.5.2.10, [7]

Table 5.3.3: Gossip times for common networks in the one–way mode

5.4   OTHER MODES AND COMPLEXITY MEASURES

In the previous sections some results and proof techniques devoted to the broadcast problem and to the gossip problem in the one–way and two–way communication modes were presented. We note that the results presented above cover only a part of the investigation of broadcasting and gossiping. The possibilities to consider complexity measures different from the number of rounds and distinct types of communication modes enable to create a lot of distinct frameworks of research problems for broadcasting, accumulating and gossiping. Some of these problem formulations may also require other considerations and proof methods for the solution of the formulated problems than the techniques presented above for the one–way mode and the two–way mode. Several of them have also direct practical applications.

The aim of this section is to give a short survey presenting informal definitions of some communication modes and some complexity measures used. We note that we are unable to present also the results connected with these modes and measures in this short survey. Anybody interested in some of these modes and/or complexity measures is refered to the corresponding literature.

Let us first start the discussion about the complexity measures. As already noted, the number of rounds as the complexity measure corresponds to the number of communication steps (i. e., to the parallel time), each of them realized in parallel. This measure is appropriate if each of the communication steps is realized approximately in the same amount of time. This can be true if the time needed to organize (synchronize) the communication step is greater than the time for direct communication (message exchange) or if in each step messages of the same length are submitted. If one has a network model in which the time for synchronization is negligible in comparison with the time for the communication, then the time intervals needed to realize distinct communication steps may be of very different lengths. To see this one can consider the broadcast (gossip) algorithm for the hypercube H_n, where in the first round each sender sends exactly one piece of information, and in the last n-th round a message consisting of 2^{n-1} pieces of information is submitted. Thus, the execution of the last round can take much longer than the execution of the first round. If, for instance, one assumes that the time needed to send a message is linearly dependent on the length of the message, then the complexity measure can be defined as follows. First, the complexity of a round is defined as the length of the longest message submitted in this round (the length of a message may be measured as the number of pieces of information included in it). Then, the final complexity of a communication algorithm corresponding to the parallel time is the sum of the complexities of all rounds. Everybody may use another modification of this measure depending on the real behaviour of his network (parallel computer). (For an overview of this model, see e.g. [18]).

Another extreme approach is not to measure the parallel time (number of rounds, etc.) but only the whole amount of exchanged messages (pieces of information) during all rounds of the communication algorithm. Such a measure may correspond to the whole communication work made by the considered network realizing a given communication algorithm. This complexity measure for communication problems was extensively investigated in the early seventies (see, for instance, [9, 25]).

To measure the efficiency of some communication algorithm under some real computing model can require to consider a trade–off between the parallel time and the work of the network. Each practical application may prefer another trade–off and we will not try to give a survey of all trade–offs considered till now.

Now, let us discuss the communication modes. All modes presented here can be considered as a generalization of the one–way communication mode. One possiblility to generalize this mode is to allow more actions for an active node in one round. (Note that an active node in the one–way communication round is either the sender or the receiver.) For instance, (i, j)-mode means that in any round one node can send a message to i neighbours via i adjacent edges and it can receive messages from j neighbours via j adjacent edges. Thus, the two–way mode is a restricted $(1, 1)$-mode, where additionally any active node must use the same adjacent edge for both submission and reception. The (i, j)-modes with several possible additional restrictions provide a rich variety of communication modes for further investigation (see, for instance, [14]).

Another possibility to generalize the one–way (two–way) mode is to consider the rounds of one–way communication algorithms as sets of vertex–disjoint paths of length one instead of as sets of directed edges (which is clearly equivalent). The generalization consists of allowing an arbitrary length of these paths in each round. Thus, a round is described by a set of vertex–disjoint paths, where additionally a direction from one endpoint to the second endpoint may be prescribed for any path. What can happen on these paths in one round is determined now by the communication mode. Obviously, this provides several possibilities. The two possibilities used in the literature [16, 17] considering either that one end–node broadcasts its whole knowledge to all other nodes of the given path or that one of the end–nodes sends its knowledge to the other end–node and the remaining nodes on the path do not read the message submitted. Especially, for the first possibility, several optimal communication algorithms [17] were constructed. Note that exactly this mode has the property that the complexity of the accumulation problem essentially differs from the complexity of the broadcast problem for several families of graphs. These modes are called vertex–disjoint modes and one possibility to generalize them is to define so called edge–disjoint modes [16], where each round is described by a set of edge–disjoint paths and what can happen on one path in one round can be determined in different ways. These modes were also investigated in [16, 17].

Obviously, one can introduce a variety of further communication modes based on other generalizations of the one-way mode. But doing this one has to be careful in order not to create an unrealistically powerful communication mode. This is not only the problem of the creation of a communication mode and a complexity measure providing some useful information about the quality of a realistic model for parallel computing, but also a problem of pure mathematical nature. Too powerful communication modes enable mostly to reach optimal communication algorithms for many graphs in a too easy way, and so the investigation of such modes does not produce any new, deep proof technique useful for other applications.

Acknowledgements

The authors would like to thank Elena Stöhr for a careful reading of the manuscript.

REFERENCES

[1] S.B. Akers, B. Krishnamurthy, "A group-theoretic model for symmetric interconnection networks", *IEEE Transactions on Computers*, Vol. 38, No. 4, pp. 555-566, 1989.

[2] F. Annexstein, M. Baumslag, A.L. Rosenberg, "Group action graphs and parallel architectures", *SIAM J. Comput.*, Vol. 19, No. 3, pp. 544-569, 1990.

[3] P. Berthomé, A. Ferreira, S. Perennes, "Optimal information dissemination in star and pancake networks", In: *Proc. 5th IEEE Symp. on Parallel and Distributed Processing (SPDP '93)*, 1993, 720–724.

[4] J.-C. Bermond, P. Hell, A.L. Liestman, J.G. Peters, "Broadcasting in bounded degree graphs", *SIAM Journal on Discrete Maths.* 5, pp. 10-24, 1992.

[5] A. Bagchi, S.L. Hakimi, J. Mitchem, E. Schmeichel, "Parallel algorithms for gossiping by mail", *Inform. Proces. Letters* 34 (1990), No. 4, pp. 197-202.

[6] C.A. Brown, P.W. Purdom, *The Analysis of Algorithms*, Holt, Rinehart and Winston, New York, 1985, § 5.3.

[7] J.-C. Bermond, C. Peyrat, "Broadcasting in deBruijn networks", Proc. 19th Southeastern Conference on Combinatorics, Graph Theory and Computing, *Congressus Numerantium* 66, pp. 283-292, 1988.

[8] J.C. Bermond, C. Peyrat, "De Bruijn and Kautz networks: a competitor for the hypercube?", In F. Andre, J.P. Verjus, editors, *Hypercube and Distributed Computers*, pp. 279-294, North-Holland, 1989.

[9] B. Baker, R. Shostak: "Gossips and Telephones", *Discr. Mathem.* 2 (1972), pp. 191-193.

[10] R.M. Capocelli, L. Gargano, U. Vaccaro, "Time Bounds for Broadcasting in Bounded Degree Graphs", *15th Int. Workshop on Graph-Theoretic Concepts in Computer Science* (WG 89), LNCS 411, pp. 19-33, 1989.

[11] G. Cybenko, D.W. Krumme, K.N. Venkataraman, "Simultaneous broadcasting in multiprocessor networks", *Proc. International Conference on Parallel Processing* (1986), pp. 555-558.

[12] G. Cybenko, D.W. Krumme, K.N. Venkataraman, "Gossiping in minimal time", *SIAM J. Comput.* 21 (1992), pp. 111-139.

[13] N.G. De Bruijn, "A combinatorial problem", *Koninklijke Nederlandsche Akademie van Wetenschappen Proc.*, Vol. 49, pp. 758-764, 1946.

[14] S. Even, B. Monien, "On the number of rounds necessary to disseminate information", *Proc. 1st ACM Symp. on Parallel Algorithms and Architectures*, Santa Fe, June 1989, pp. 318-327.

[15] R.C. Entringer, P.J. Slater, "Gossips and telegraphs", *J. Franklin Institute* 307 (1979), pp. 353-360.

[16] A.M. Farley, "Minimum-Time Line Broadcast Networks", *Networks*, Vol. 10 (1980), pp. 59-70.

[17] R. Feldmann, J. Hromkovič, S. Madhavapeddy,B. Monien, P. Mysliwietz, "Optimal algorithms for dissemination of information in generalized communication modes", *Proc. PARLE'92*, Lecture Notes in Computer Science 605, Springer Verlag 1992, pp. 115-130.

[18] P. Fraigniaud, E. Lazard, "Methods and problems of communication in usual networks", *Discrete Applied Mathematics* 53 (1994), No. 1-3, 79–133.

[19] A.M. Farley, A. Proskurowski, "Gossiping in grid graphs", *J. Combin. Inform. System Sci.* 5 (1980), pp. 161-172.

[20] R. Feldmann, W. Unger, "The Cube-Connected Cycles Network is a Subgraph of the Butterfly Network", *Parallel Processing Letters*, Vol. 2, No. 1 (1992), pp. 13-19.

[21] C. Gowrisankaran, "Broadcasting in recursively decomposable Cayley graphs", *Discrete Applied Mathematics* 53 (1994), No. 1-3, 171–182.

[22] S.M. Hedetniemi, S.T. Hedetniemi, A.L. Liestman, "A survey of gossiping and broadcasting in communication networks", *Networks*, Vol. 18, pp. 319-349, 1988.

[23] J. Hromkovič, C. D. Jeschke, B. Monien, "Optimal algorithms for dissemination of information in some interconnection networks", extended abstract in *Proc. MFCS'90*, Lecture Notes in Computer Science 452, Springer Verlag 1990, pp. 337-346; full version in: *Algorithmica* 10 (1993), 24–40.

[24] J. Hromkovič, C.D. Jeschke, B. Monien, "Optimal algorithms for dissemination of information in some interconnection networks", *Theoretical Computer Science* 127 (1994), No. 2, 395–402.

[25] A. Hajnal, E.C. Milner, E. Szemeredi, "A cure for the telephone disease", *Can. Math. Bull.* 15 (1972), pp. 447-450.

[26] M.C. Heydemann, J. Opatrny, D. Sotteau, "Broadcasting and spanning trees in de Bruijn and Kautz networks", *Discrete Applied Mathematics* 37/38 (1992), 297–317.

[27] R. Klasing, R. Lüling, B. Monien, "Compressing cube-connected cycles and butterfly networks", extended abstract in *Proc. 2nd IEEE Symposium on Parallel and Distributed Processing*, pp. 858-865, 1990; full version in *Discrete Applied Mathematics* 53 (1994), No. 1-3, 183-197.

[28] R. Klasing, B. Monien, R. Peine, E. Stöhr, "Broadcasting in Butterfly and DeBruijn networks", *Proc. STACS'92*, Lecture Notes in Computer Science 577, Springer Verlag 1992, pp. 351-362.

[29] D.E. Knuth, *The art of computer programming*, Vol. 1, Addison-Wesley, Reading-Massachusetts, 1968.

[30] W. Knödel, "New gossips and telephones", *Discrete Math.* 13 (1975), p. 95.

[31] D.W. Krumme, "Fast gossiping for the hypercube", *SIAM J. Comput.* 21 (1992), to appear.

[32] A.L. Liestman, J.G. Peters, "Broadcast networks of bounded degree", *SIAM Journal on Discrete Maths*, Vol. 1, No. 4, pp. 531-540, 1988.

[33] R. Labahn, I. Warnke, "Quick gossiping by multi-telegraphs", In R. Bodendiek, R. Henn (Eds.), *Topics in Combinatorics and Graph Theory*, pp. 451-458, Physica-Verlag Heidelberg, 1990.

[34] B. Monien, I.H. Sudborough, "Embedding one Interconnection Network in Another", *Computing Supplementum* 7 (1990), 257-282.

[35] V.E. Mendia, D. Sarkar, "Optimal broadcasting on the star graph", *IEEE Transactions on Parallel and Distributed Systems* 3 (1992), No. 4, 389–396.

[36] M.G. Pease, "An adaptation of the Fast Fourier transform for parallel processing", *Journal ACM*, Vol. 15, pp. 252-264, April 1968.

[37] D. Richards, A.L. Liestman, "Generalization of broadcasting and gossiping", *Networks* 18 (1988), pp. 125-138.

[38] M.R. Samatham, D.K. Pradhan, "The de Bruijn multiprocessor network: a versatile parallel processing and sorting network for VLSI", *IEEE Transactions on Computers*, Vol. 38, No. 4, pp. 567-581, 1989.

[39] E.A. Stöhr, "Broadcasting in the butterfly network", *Information Processing Letters* 39 (1991), pp. 41-43.

[40] E.A. Stöhr, "On the broadcast time of the butterfly network", *Proc. 17th Int. Workshop on Graph-Theoretic Concepts in Computer Science* (WG 91), LNCS 570, pp. 226-229, 1991.

[41] E.A. Stöhr, "An upper bound for broadcasting in the butterfly network", *manuscript*, University of Manchester, United Kingdom.

[42] V.S. Sunderam, P. Winkler, "Fast information sharing in a distributed system", *Discrete Applied Mathematics* 42 (1993), pp. 75-86.

[43] I.H. Wilkinson, *The algebraic eigenvalue problem*, Clarendon Press, Oxford, 1965.

Applied Optimization

KLUWER ACADEMIC PUBLISHERS – DORDRECHT / BOSTON / LONDON